Handbook of
COMMERCIAL SCIENTIFIC INSTRUMENTS

Volume 2
THERMOANALYTICAL TECHNIQUES

HANDBOOK OF COMMERCIAL SCIENTIFIC INSTRUMENTS

SERIES EDITORS

Claude Veillon and W. W. Wendlandt

University of Houston
Houston, Texas

VOLUME 1 ATOMIC ABSORPTION
Claude Veillon

VOLUME 2 THERMOANALYTICAL TECHNIQUES
W. W. Wendlandt

Other volumes in preparation

Handbook of COMMERCIAL SCIENTIFIC INSTRUMENTS

Volume 2

THERMOANALYTICAL TECHNIQUES

W. W. Wendlandt

Department of Chemistry
University of Houston
Houston, Texas

MARCEL DEKKER, INC. New York 1974

MARCEL DEKKER, INC.

270 Madison Avenue, New York, New York 10016

LIBRARY OF CONGRESS CATALOG CARD NUMBER: 72-87851

ISBN: 0-8247-6060-3

Current printing (last digit):
10 9 8 7 6 5 4 3 2 1

PRINTED IN THE UNITED STATES OF AMERICA

CONTENTS

INTRODUCTION TO THE SERIES

Many persons and organizations, when contemplating the purchase of a scientific instrument, are often unaware of just what is available. They frequently do not have the time or the inclination to assemble information on all of the available instruments for a particular purpose. For example, a production laboratory might have a troublesome trace metal analysis problem, for which their present analytical methods are too time consuming, inaccurate, or not sensitive enough. They therefore wish to consider atomic absorption spectrometry. Now come the questions: How much will it cost? What will it do? What are the specifications of available instruments? What accessories are available? Could we get by with an atomic absorption attachment on our Model X spectrometer? . . . and so on.

For this exemplary organization, it could be very difficult to accumulate all the information on all the commercially available atomic absorption instruments or accessories. Consequently, they might consider only two or three instruments, buy one of these, and later discover that another instrument (of which they were not initially aware) could do their particular job better.

To alleviate this rather commonplace problem, the authors of the *Handbook of Commercial Scientific Instruments* will present information furnished by all known manufacturers of selected groups of instruments. The *Handbook* will be a multi-volume series devoted to commercially available scientific instruments of various types. Each volume will fully describe the instrumentation available for a particular field. Specifications, descriptions, schematic drawings, photographs, approximate prices, accessories, etc., of each type of instrument will be included in the Handbook volumes. Material presented will be essentially that furnished by the manufacturers, plus some evaluative, objective comparisons of the instruments by the volume authors.

All of the volume authors are actively engaged in research in their respective areas and are familiar with the available instrumentation. None represents any instrument manufacturer.

Specifications, descriptions, etc., of similar instruments will be essentially those supplied by the individual manufacturers. Of course, some editing may be necessary, perhaps to put an important specification for several similar instruments on the same basis. For example, resolution is defined in several ways by the various manufacturers of mass spectrometers. These might be recalculated and all defined in a consistent manner, so that the user of the Handbook volumes can more conveniently evaluate and compare the various instruments.

The authors hope to include in each volume all instruments of a given type that are commercially available in the United States, regardless of their national origin.

The first volume deals with commercially available atomic absorption instruments. This volume describes commercially available thermal analysis instrumentation. Subsequent volumes will deal with nuclear magnetic resonance instruments, gas chromatographs, mass spectrometers, x-ray instruments, spectrophotometers, various types of electronic equipment, and many others.

The editors sincerely believe that the *Handbook of Commercial Scientific Instruments* will fulfill a definite need, a need very often expressed as a question by individuals and organizations -- which one should we buy? We feel that if the reader knows what instruments are available and what they will do, then the correct decision can be made.

CLAUDE VEILLON

W. W. WENDLANDT

PREFACE

The thermal analysis instruments described in the volume include only those for thermogravimetry (TG), differential thermal analysis (DTA), and differential scanning calorimetry (DSC). As was discussed in Volume 1 of this series, only those instruments are described that are commercially available to users in the United States. It has proved impossible to obtain the manufacturers literature on instruments produced in the USSR and by some of the Japanese companies.

Since this volume was begun, a number of manufacturers ceased to exist or else cut their number of models down to a bare minimum. In many cases, descriptions of the instruments were published by the manufacturers, but no production models were ever built. When the latter has occurred, if known, these instruments have not been included. However, no responsibility can be assumed by the author or publisher for the availability of the instruments described in this volume.

Each instrument is described as completely as space permits using data and specifications provided by the manufacturer. It was not possible to summarize the specifications of every unit into a table because of the lack of these data or the many combinations of furnaces and sample holders that are available. The reader can glean this information from the Specifications presented for each model.

The purchase of a thermobalance, DTA, or DSC instrument represents a large investment and should be considered in great detail. It is hoped that the material described herein will prove useful to the reader for making an intelligent decision depending upon his research or production needs.

It is a pleasure to acknowledge the assistance of the many manufacturers who furnished the material so necessary to make this volume possible. Also, the typing ability of Ms. Julie Norris is gratefully acknowledged.

W. W. WENDLANDT

Houston, Texas

Part I

DIFFERENTIAL THERMAL ANALYSIS (DTA)

INTRODUCTION

The technique of differential thermal analysis (DTA) consists of measuring the difference in temperature between the sample and a reference material (T_s-T_r) as both are heated or cooled at some fixed rate. The sample and reference materials are contained in a furnace the rate of heating or cooling of which is controlled by a temperature programmer. The technqiue only records sample transitions caused by enthalpy changes (except for second-order transitions) in the temperature region of interest. Such enthalpy changes as phase transitions-fusion, solid-solid, boiling, sublimation, and decomposition, oxidation-reduction, and so on, can be detected. The qualitative aspects of the technique consist of determining the number and magnitude of the curve peaks as a function of temperature. Evaluation of the area of the peaks constitutes the quantitative aspect of the technique, since the peak area is related to the ΔH of the transition.

A schematic diagram of a DTA instrument is given in Fig. 1.

The sample and reference materials are contained in various types of holders, such as cavities in a metal or ceramic block; metal cups of different configurations; glass, quartz, or ceramic tubes, and so on. Both the sample and reference are in intimate contact with a temperature sensor, which is generally a thermocouple, although resistance thermometers, thermistors, thermopiles, and thermoelectric disks are also used. The most common thermocouple is that

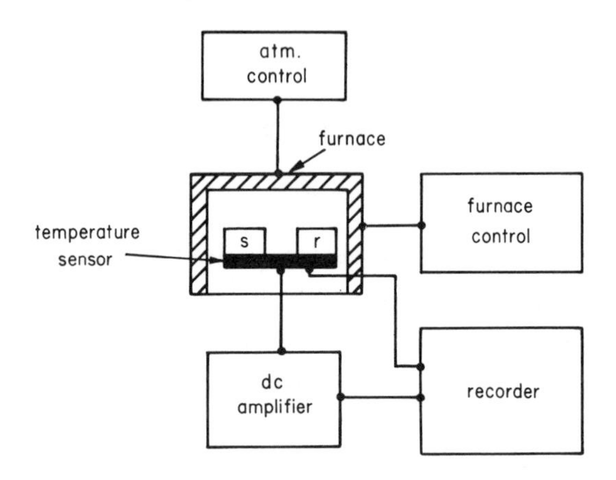

FIG. 1. Schematic diagram of DTA instrument.

composed of Chromel-Alumel wire. Chromel-Constantan, Platinel II, and Pt, Pt-10%Rh are also employed, depending upon the temperature range. Photodiodes have also been used as temperature sensors, but they are not being used as such in any commercially available instrument.

Many types of furnace are available in commercial DTA instruments. The most common appears to be a Kanthal wire-wound ceramic core tube furnace for use up to about 1250°C. To attain higher temperatures, platinum, rhodium, or silicon carbide heating elements are employed. Temperature increase or decrease of the furnace is controlled by the temperature programmer. Many different programmers are available, ranging from simple motor-advanced variable voltage transformers to thermocouple-actuated feedback-type controllers. Heating or cooling rates are generally variable from 1° to 20°C/min; the most commonly used rate is 10°C/min.

The temperature difference between the sample and reference materials ($T_s - T_r$) is amplified by a low-noise, stable dc amplifier (if thermocouples are used as temperature sensors) and recorded on a recorder. The recorder may be a two-channel potentiometric strip-chart type, in which one channel is used to record the T_r-T_s signal and the other, for the sample, reference, or furnace temperature. Both functions are, of course, recorded against time. An X-Y recorder in which the T_s-T_r signal is plotted as a function of temperature (T_s, T_r, or T_f) may also be used. For calorimetric use, the strip-chart recorder is preferred; the X-Y recorder is used for greater temperature accuracy. Generally, both types of recorder are available from the manufacturer.

The DTA curve is highly dependent on the nature of the gaseous atmosphere surrounding the sample; hence, for consistent results, the atmosphere must be rigorously controlled. Furnace atmospheres may be static or dynamic, in inert or reactive gases, and at high or low pressures. Commonly used pressures vary from 10^{-5} Torr to 3000 psig. Some furnaces have provisions for the use of corrosive gases or vapors of low-boiling liquids (water and organic liquids).

Difference between DTA and DSC

In recent years the nomenclature in DTA has become rather confused. This is especially true in the case of the terms "differential scanning calorimetry," "differential thermal analysis," "differential enthalpic analysis," "dynamic differential calorimetry," "quantitative differential thermal analysis," and so on. "Differential scanning calorimetry" was coined by the Perkin-Elmer Corporation to describe a new approach to differential thermal measurements, quantitative in nature, which was embodied in their commercial instruments, the P-E DSC-1 and DSC-1B.

The term DSC was introduced to distinguish the Perkin-Elmer calorimeter from other techniques for determining differential thermal measurements. The distinction is based on the "null-balance" principle of measurement, in which heat energy absorbed or evolved by the sample is compensated for by adding or removing an equivalent amount of electrical energy to a heater located in the sample holder. In practice this is achieved by comparing the signal from a platinum resistance thermometer in the sample holder with that from an identical sensor in a reference holder. The continuous and automatic adjustment of heater power (energy per unit time) necessary to keep that of the sample holder identical to that of the reference holder provides a varying electrical signal, opposite but equivalent to the varying thermal behavior of the sample. By recording the variable part of the electrical signal or differential power, a record of the sample behavior is obtained in power units and expressed in millicalories per second. The peak area is a true electrical energy measurement, the magnitude of which does not depend on any of the thermal constants of the sample or apparatus.

The calibration constant, E, to convert measured DSC data to calories or millicalories, is independent of temperature. In the Perkin-Elmer DSC-1B instrument, this is true to within ±2% over any 500°C range. In DTA, however, the E value changes with temperature and is dependent upon the thermal parameters of the various parts of the apparatus and the thermoelectric power of the thermocouple. It may vary as much as a factor of three over the same 500°C range. Thus, multipoint calibrations must be used in DTA; and broad peaks must be divided up and measured in sections, each section having its own average E value.

In classical DTA, the quantity recorded as the ordinate is almost always T_s-T_r (ΔT) or the reverse, T_r-T_s. The magnitude of ΔT for a given rate of heat energy change is directly proportional to the effective thermal resistance of the system. The major source of resistance is between the source of heat and the point where the temperature is sensed. In classical

DTA, therefore, the sample itself forms a major part of the conduction path; hence, the block-type holder is practically useless for quantitative calorimetric measurements.

To improve the calorimetric performance of DTA systems, Boersma (in 1955) found that it was necessary to remove the temperature sensor from the body of the sample, thus eliminating from the E value the unknown and variable contribution of the sample's thermal resistance. His apparatus removed the sample dependence of E, but the temperature dependence remained. Thus, to make a reliable calorimetric measurement, calibration with a standard having a transition temperature close to that of the unknown is required.

One manufacturer, the Du Pont Company, refers to their Boersma-type DTA cell as a "differential scanning calorimeter." A more reasonable term, to distinguish it from true DSC, is perhaps "differential enthalpic analysis" (DEA) or "dynamic differential calorimetry" (DDC). It is not always possible, by an inspection of an unlabeled thermal analysis peak, to determine which type of apparatus was used. The differences in basic measuring principle show up very clearly, however, in the instrument calibration and data analysis.

Commercial DTA Instruments

A wide variety of DTA instruments are available from the manufacturers. The choice of instrument is dictated by the nature of the application, the future application, the budget, and the laboratory space available. Many manufacturers offer complete thermal analysis systems, which generally include DTA, thermogravimetry, evolved gas detection, dilatometry, and so on. Each technique is in a modular form so that components may be added at any time. Generally, the basic recording and controller unit must be purchased and the DTA and TG modules added to it. Also, there is a wide choice of DTA modules, depending upon the nature of the sample holder and the temperature range of interest. High-temperature furnaces are naturally more expensive than lower range models, due to the expensive platinum alloy windings.

The instrumental descriptions which follow were taken from the manufacturer's literature. In many cases, vital statistics were not available or were not given in the literature. An attempt was made to include all of the instruments that are currently available to workers in the United States.

A.D.A.M.E.L.

ATD-67 Differential Thermal Analyzer Models 1 and 2

The ATD-67 DTA instrument, illustrated in Fig. 2, consists of three components. The middle instrument case contains the recorder, two amplifiers (one for measuring the temperature and the other for the differential temperature), and part of the furnace temperature programmer. On either side of the control and recording console are the two furnace and sample holder assemblies. The latter may be of two types: Model 1 is capable of an air or inert gas atmosphere, while Model 2 is used for static controlled atmospheres or at reduced pressures.

The recorder is a two-pen galvanometer strip-chart type, in which one pen is used to record the sample or reference temperature while the other records the ΔT signal. There are three chart speeds on the recorder - 20, 60, and 120 mm/hr.

Furnaces of two types are available, either of which may be mounted on the Model 1 or 2. The first has a maximum temperature of 1050°C, while the second has a maximum temperature limit of 1200°C. A temperature cycle fixed on a drum rotating at a constant speed enables different heating and cooling rates to be obtained. As normally supplied, the drum is driven by a clock-motor with two speeds obtained by changing gears. The two cycles supplied correspond to heating rates of 300° and 600°C/hr.

The convenience of two furnaces and sample holders is that while one furnace is cooling, the other may be employed in the heating cycle.

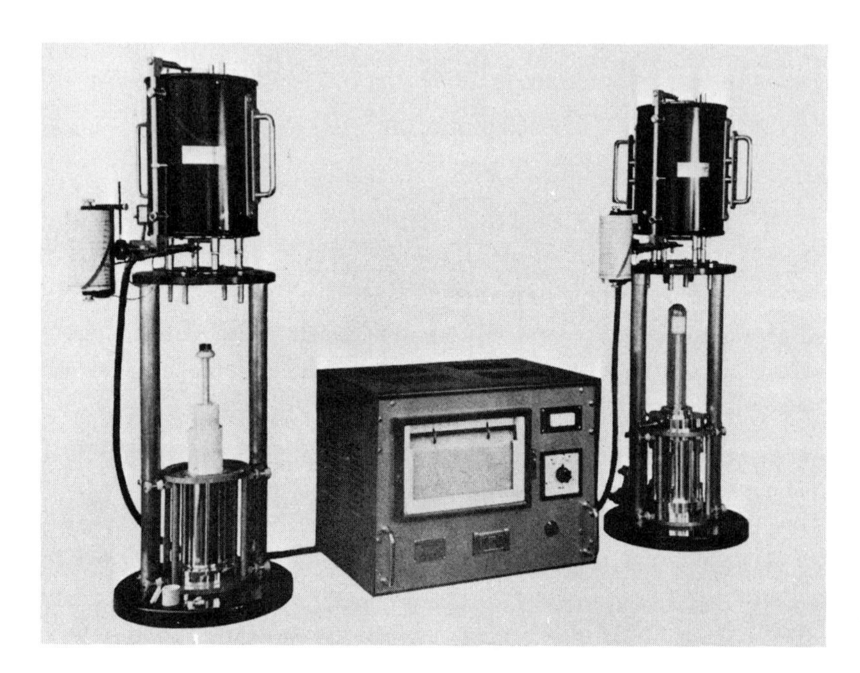

FIG. 2. A.D.A.M.E.L. ATD-67 DTA instrument.

SPECIFICATIONS

Control Console

Recorder: X_1-X_2-time galvanometer type. X_1-axis, 100 mm wide, three scales corresponding to 0°-500°C, 0°-1200°C, and 0°-1500°C.

ΔT: X_2-axis, 100 mm wide, continuously variable from 100 to 200 μV. Zero can be shifted continuously from ±40%.

Chart-speed: 20, 60, and 120 mm/hr.

Case dimensions: 490 x 385 x 380 mm.

Electrical requirements: 220 V, 50 H, which can be converted to 127 V with supplied transformer. Total power consumption is 1500 W.

Furnace Assemblies

Thermocouple composition: Pt, Pt-10%Rh; Chromel-Alumel may also be supplied for use below 1000°C.

Dimensions:

Model 1: 330 mm in diameter by 950 mm in height.

Model 2: 330 mm in diameter by 950 mm in height.

ATD-63 Differential Thermal Analyzer, Models 1 and 2

The ATD-63 Differential Thermal Analyzer is similar to the ATD-67, with a more sophisticated recorder, sample holder, and furnace. The difference between the Models 1 and 2 is that the latter can be used for studies at low pressures, while the former permits only purging with an inert gas. The ATD-63 is shown in Fig. 3.

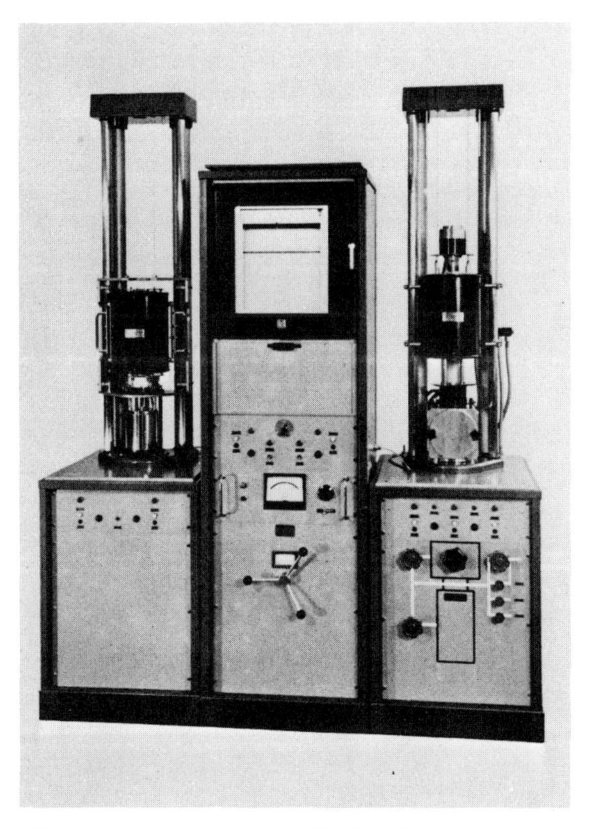

FIG. 3. A.D.A.M.E.L. ATD-63 DTA instrument.

SPECIFICATIONS

Recorder: X_1-X_2-time potentiometric, Speedomax X_1X_2T.

X_1-axis: 1, 2.5, 5, 10, and 25 mV; zero can be continuously moved 0 to 300% of the scale.

X_2-axis: For a Pt, Pt-10%Rh thermocouple, 0°-1000°C and 0°-1500°C.

Chart speeds: 50, 100, 150, 200, 300, and 400 mm/hr.

ΔT amplifier: 50, 100, 200, 500, 1000, and 2000 μV full scale.

Heating rates: 300° and 600°C/hr.

Metal case dimensions: 1820 x 620 x 560 mm.

Model 1 case: 2160 x 620 x 560 mm.

Model 2 case: 2160 x 620 x 560 mm. Consists of (a) for studies under reduced pressures or controlled atmosphere; (b) for use under high vacuum. The latter contains a high vacuum system.

AMERICAN INSTRUMENT COMPANY

Aminco Thermoanalyzer

The Aminco Thermoanalyzer is shown in Fig. 4. The thermoanalyzer consists of five basic components: 1) sample holder and thermocouple assembly, 2) furnace assembly, 3) flow control system, 4) differential thermocouple amplifier, and 5) furnace power programmer.

The sample holder consists of a metal block containing sample and reference materials cavities; each cavity is fitted with a thermocouple. A third thermocouple measures the block temperature. Furnace design is based upon that described by Lodding and Hammel [*Rev. Scient. Instr.*, **30**, 885 (1959)]. Maximum temperature limits are 1000°C or 1500°C.

The flow control system provides a means of maintaining a uniform atmosphere in the furnace chamber. Gas flow moves upward through the furnace chamber and into the sample and reference cavities in the sample block. The gas passes down through the sample and reference materials, down the bores of the ceramic tubes which encase the thermocouples, and emerges from the two outlets at the bottom of the furnace assembly. These outlets feed to separate flowmeters and needle valves for the sample and reference streams. The streams then flow to two outlets on the front panel, from which they can be either exhausted or piped through a gas sampling valve to other instruments (such as a gas chromatograph).

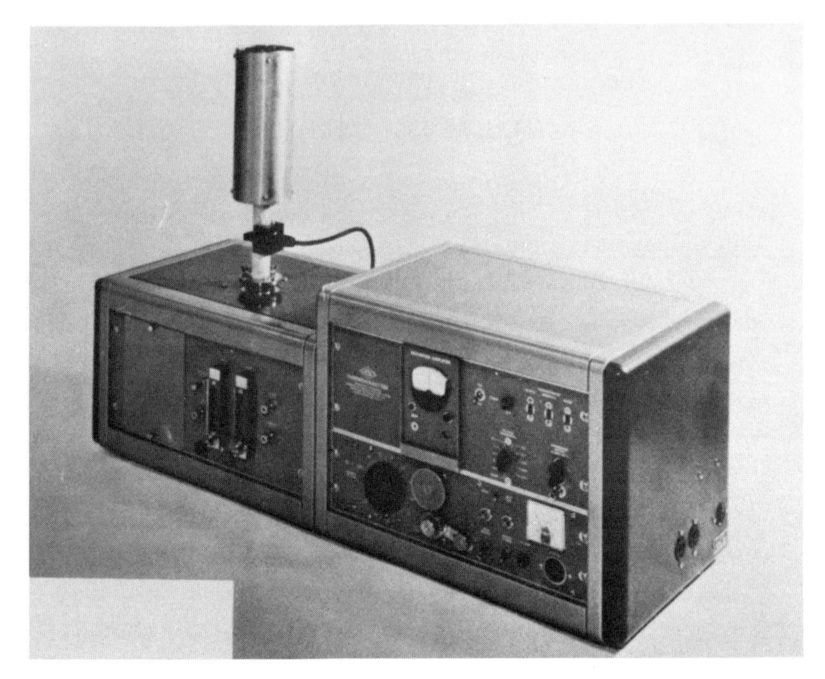

FIG. 4. Aminco thermoanalyzer.

SPECIFICATIONS

Temperature range: -75° to 1500°C by use of accessories.

Heating rates: 1/4, 1/2, 1, 2, 4, 8, 16, 18, 21, and 24°C/min.

Sample size: 0.5 to 400 mg.

DC amplifier: Low-noise, chopper input.

Accessories:

Effluent Gas Analysis Accessory: Consists of a two-thermistor detector, Wheatstone bridge, and power supply.

Gas Sampling Valve: For connection of thermoanalyzer to a gas chromatograph or other instrument.

Low Temperature Accessory: Consists of a Fiberfax sleeve which is placed over the top of the standard furnace, inverted, and cooled with liquid nitrogen. Permits studies to -75°C.

High Temperature Accessory: For temperatures up to 1500°C, Pt, Pt-Rh thermocouples replace the Chromel-Alumel type, while the stainless steel block is replaced by a ceramic one.

COLUMBIA SCIENTIFIC INDUSTRIES

Stone Analytical Instrument 200 Series

The Stone 200 Series Systems use a modular approach. For a DTA-DSC system, the components consist of a recorder-controller module, a furnace platform, a furnace, and a sample holder. Several choices of furnaces are available as well as numerous sample holders.

A typical 200 Series system is shown in Figure 5. A unique feature of these instruments is the dynamic gas atmosphere

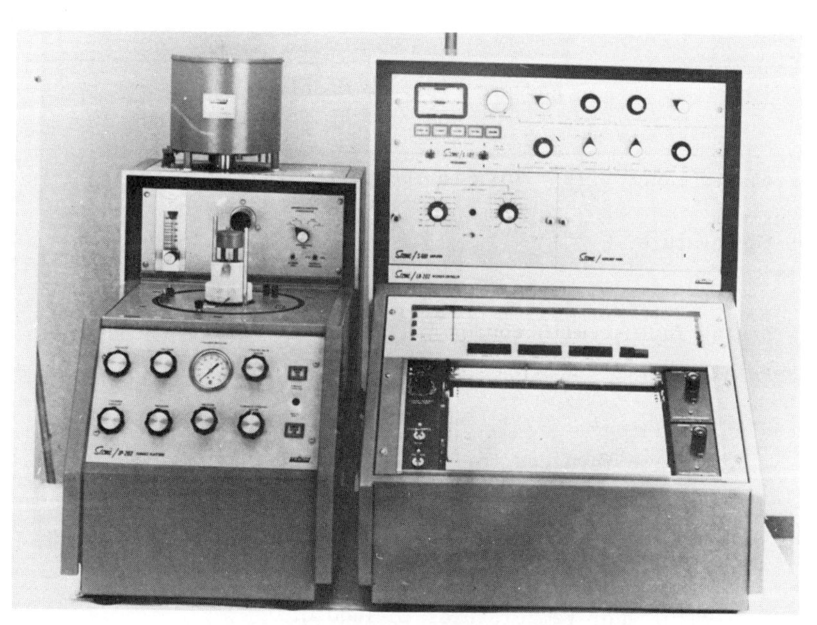

FIG. 5. Stone DTA system.

control. Not only can the dynamic atmosphere consist of permanent-type gases, but steam or low-boiling organic compounds may also be employed. A dynamic steam atmosphere can reveal new information in studies in which water plays a significant role. It may also displace transition temperatures, allowing clarification of curve peaks by comparison with curves in other atmospheres. The hydrolytic stability of a polymer or the plasticizing effects of moisture may also be determined.

There are four models in the 200 Series from which to choose, the various components of which are shown in Table 1. They differ in the type of furnace, recorder, sample holder, and so on. Each component will be discussed subsequently.

There are two furnace platforms, the specifications of which are given in the following tabulation.

	Furnace Platform Model	
	P-202	P-202C
Pressure range	10^{-2} Torr to 100 psig	10^{-2} Torr to 250 psig
Maximum temperature	1200° or 1600°C	500°C

Note. Special 202 Models with vapor generators and heater wrapping for use with condensible gases are available on special order.

The three furnaces for the 200 Series instruments are:

F-202RC Quick cool for temperatures to 700°C, which can be air-cooled to 25°C from 500°C in 15 min.

F-202LT For temperatures to 1100°C.

F-202HT For temperatures to 1600°C.

TABLE 1

Stone 200 Series Systems

Model 210	Complete 0° to 1200°C DTA system. Includes RC-202A recorder controller; X-Y recorder; P-202 furnace platform; F-202LT (0° to 1200°C) furnace; sample holder with either ring, exposed, or post-type differential thermocouples; and all accessories.
Model 211	Complete 0° to 1600°C DTA system. Includes RC-202A recorder controller, X-Y recorder, P-202 furnace platform, F-202HT (0° to 1600°C) furnace, SH-202-13-PT sample holder, and all accessories.
Model 213	Complete 150° to 500°C subambient DTA-DSC system. Includes RC-202G recorder controller with dual pen strip chart recorder, P-202 furnace platform, DSC-200-5 subambient DSC accessory, and accessory kit.
Model 214	Complete 0° to 1000°C DSC-DTA system. Includes RC-202G recorder controller with dual pen strip chart recorder, P-202 furnace platform, DSC-202-2 sample holder, F-202LT (0° to 1200°C) furnace, and all accessories.

The Recorder-Controller for the Stone system may be used for operation of the DTA, TG, or x-ray diffractometer furnaces. A choice of recorders, either X-Y or strip-chart, is available, as well as a wide variety of heating rates and other control functions. The specifications of the 202 Series Recorder-Controller are shown below.

SPECIFICATIONS

Programmer

Rates: 0.5° to 50°C/min, infinitely adjustable, plus ten highly reproducible set rates.

Linearity: Better than 0.25% for any 100°C interval, or better than 0.5% full scale.

Limit switch modes: Off, Standby, Hold, Heat (Cool), Cycle between infinitely adjustable upper and lower limits. Limits accurate to within 0.5%.

Amplifier/Recorder

Sensitivity: Infinitely adjustable to give full-scale output from inputs of 0.005 to 4 mV. Maximum of 0.002°C per chart division with Platinel thermocouple and aluminum sample holder.

Noise: Less than 0.2% of full scale at maximum amplification.

T-scales for models with X-Y recorder: Rotary switch-selected, recorder full-scale spans for -190° to 150°C, -190° to 500°C, 0° to 250°C, 0° to 270°C, 0° to 500°C, 0° to 550°C, 0° to 1000°C, 0° to 1100°C, and 0° to 1600°C, according to thermocouple used.

T-scales for models with strip-chart recorder: Push-button selected, full-scale readout spans for -200° to 120°C, -200° to 350°C, 0° to 275°C, 0° to 550°C, 0° to 1100°C and 0° to 1600°C, according to thermocouple used.

Chart speeds for strip-chart recorders: 1 and 2 in./hr; 0.1, 0.2, 0.5, 1, and 2 in./min; 0.1, 0.2, 0.5, 1, and 2 in./sec.

Dimensions and weight: 21 x 25 x 25 inches; 90 lb.

Standard Models

RC-202A: Recorder-Controller - single-channel, including programmer, amplifier, and single-pen X-Y recorder. (Can be furnished without amplifier for TG.)

RC-202C: Same as RC-202A except with two-pen $X\text{-}YY_1$ recorder for two-channel operation.

RC-202F: Same as RC-202A, but with strip-chart recorder with two different pens and event-marker pen. Visual temperature readout for various spans plus atomatic event marker at 10°, 25°, or 50°C intervals, depending on span used.

Three different types of sample holders are available for the Stone DTA systems, differing mainly in their sample container configuration. The sample may be contained in a cylindrical cavity in a metal block, a shallow dish, or a cylindrical cup or crucible.

The DSC-202 Series sample holder is shown in Fig. 6. This holder contains an exposed-loop differential thermocouple with direct contact with the sample. It features full dynamic gas flow and is used for soil, mineral, and other materials which do not melt or sinter.

The ring-thermocouple holder is illustrated in Fig. 6. The sample is contained in a shallow metal dish which is exposed to the dynamic gas atmosphere. Sample sizes may vary from 0.1 to 20 mg.

A less-sensitive sample holder is that shown in Fig. 6. Samples are contained in cylindrical metal crucibles, which permit larger sample sizes, ranging from 20 to 200 mg. There is, of course, less exposure of the sample to the dynamic gas flow.

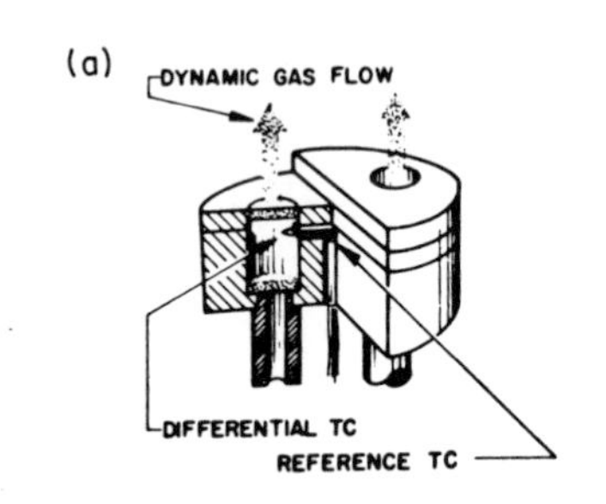

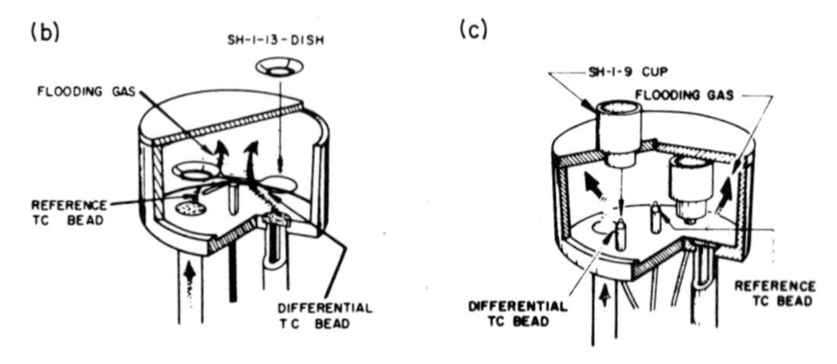

FIG. 6. Sample holder configurations: (A) SH-8BE2 series; (B) SH-11BR2(4) series; (C) SH-11BP2(4) series.

Sample holder composition and the types of thermocouples are described below.

SPECIFICATIONS

Sample Holders

Aluminum (Al): Maximum 550°C. Excellent stability with negligible thermal drift. Very suitable for plastics.

Nickel (Ni): Maximum 1100°C in inert gas, 700°C with O_2 dynamic gas. Do not use with high levels of Cr, V, Pb, or S.

Stainless steel (SS): Maximum 1000°C in inert gas, 950°C with air dynamic gas, 700°C with O_2 dynamic gas. More resistant than Ni and suitable for Cr, V, Pb, and S.

Palladium-Ruthenium (PD): Maximum 1370°C. Do not use

with H_2 or H_2S dynamic gas. Produces weak extraneous peak from 800° to 830°C; do not use for measurement in this case.

Platinum - 10° Rhodium (PT): Maximum 1600°C. Do not use with H_2 or H_2S dynamic gas.

Thermocouples

Iron-Constantan (J): For full subambient range.

Chromel-Alumel (K): Maximum 1100°C in inert gas; 1100°C in reducing gas; 700°C in oxidizing gas.

Platinel II (P): Maximum 1100°C. Do not use with H_2 or H_2S dynamic gas. Up to six times better sensitivity than S types.

Platinum - Platinum 10% Rhodium (S): Maximum 1600°C. Do not use with H_2 or H_2S dynamic gas.

A high-pressure sample holder for pressures to 3000 psig is also available as an optional accessory. The SH-15 BR2-SS sample holder consists of an enclosed ring-type thermocouple sample holder, in which the applied gas pressure can be varied from atmospheric to 3000 psig by use of an external nitrogen gas source. Maximum pressure is derated with increase in temperature; at 500°C, the maximum allowable pressure is 2000 psig.

SPECIFICATIONS

Maximum pressure and temperature: Varies from 4000 psig at 100°C to 500 psig at 700°C. Maximum temperature is 800°C. Drift is ±0.03°C in N_2 atmosphere from 25 to 500°C at a heating rate of 10°C/min at 2000 psig.

Sensitivity: ±0.002°C per chart division.

Thermocouples: Ring-type.

Safety valve: Rupture disk set at 4000 psig.

Construction: Stainless steel block and base.

Also available is the Model SH-14BR2 High-Vacuum Sample Holder for use to pressures of 1×10^{-6} Torr at temperatures up to 1000°C.

Stone Model 500 DTA-DSC System

The Model 500 DTA and DSC systems are designed to be used in colleges and universities. These systems will give results comparable to higher priced instruments and will withstand the rigors of everyday student laboratory use. A full line of TG, DSC, and DTA sample holder modules permit the capability for expansion of the system as the research or teaching needs vary. The instrument can also be used in many industrial plants and laboratories for quality control and as a second instrument where high performance research instruments are overloaded with thermal analysis work.

The specifications for the basic unit (RCP-500-V), used in all Model 500 Series Systems, are as follows.

SPECIFICATIONS

Differential Amplifier: Plug-in modular construction, solid state, chopper stabilized. Signal to noise ratio 300 to 1 on most sensitive range. Eight selectable ranges of 10, 20, 40, 80, 100, 500, and 1000 mV. Amplifier output for ΔT recorder is 10 mV full scale on all ranges. Will drive any 10 mV recorder having impedance greater than 1000 ohms.

Programmer: Plug-in modular construction, solid state. Switch-selected rates of 1°, 2°, 5°, 10°, 20°, and 50°C/min. Linearity better than 1% full scale.

Programmer Modes: Off, Standby, Manual, and Automatic at selected rate. Programs up and down with subambient accessories.

Limit Switch Function: May be set at any point between temperature limits of furnace and sample holder.

Power: 115 V, 60 Hz, 15 A.

Size: Overall dimensions are 21 in. long, 10 in. high,

18 in. deep.

Weight: Approximate shipping weight is 60 lb.

There are four Models in this series, each has a different furnace, sample holder, recording system, and so on. The components which make up each Model are:

Model 509 -- Complete 0° to 1200°C DTA System without recorder. Includes RCP-500-V, F-500 (0° to 1200°C) furnace, SH-500 (0° to 1200°C) sample holder, either exposed or ring type TC, and all accessories.

Sample size -- less than 0.01 mg to over 200 mg.

Atmosphere -- Ambient to 10^{-2} mm Hg and gas sweep.

Model 510 -- Complete 0° to 1200°C DTA System with X-Y recorder. Includes Model 509 and all accessories.

Model 511 -- Complete -150° to +600°C DTA System with X-Y recorder. Includes RCP-500-V, SA-500-55 subambient accessory, and all system accessories.

Programming -- Programs both down and up to temperature limits.

Sample Size -- Less than 1 mg to 200 mg.

Atmosphere -- Ambient; provision for gas sweep.

Model 512 -- Complete -150° to +600°C DSC-DTA System with dual channel strip chart recorder. Includes RCP-500-V, DSC-202-55 subambient DSC accessory, and all other system accessories.

Sensitivity -- 2 mcal/sec. full scale.

Sample Size -- Less than 1 mg to 200 mg.

Atmosphere -- Ambient; provision for gas sweep.

DELTATHERM

Two series of Deltatherm instruments are available: the Deltatherm II, a complete research instrument; and the Deltatherm III, a new low-cost, all-purpose unit. Although the Deltatherm II is primarily a research instrument, its four-sample DTA holder can also be used to great advantage in process and quality control, where the number of analyses per day or hour is a critical factor. The Deltatherm III was developed to meet the growing needs of both process control and educational laboratories.

Deltatherm II

This is a basic research system with integrated modular accessories built around the D2000 basic unit. Accessories permit the determination of DTA curves from -200° to 1600°C on one to four samples simultaneously; thermogravimetry on samples from several milligrams to 100 g to 1600°C; dynamic adiabatic calorimetry of ±2% accuracy to 800°C; simultaneous TG-DTA; derivative plotting; and digital readout for computer processing.

The basic unit, as shown in Fig. 7, is the controlling and recording system and provides (1) temperature programming with ±2% linearity at 100 or more discrete heating rates between 0° and 20°C/min., (2) a four-channel recorder which uses electrosensitive paper to record the four simultaneous curves, and (3) control panel and indicator meters.

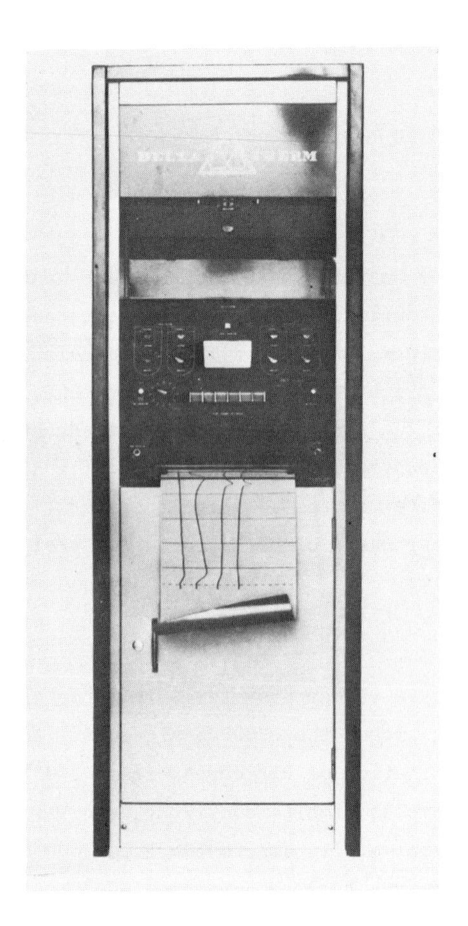

FIG. 7. The Deltatherm II basic unit.

Differential thermal analysis measurements are obtained by use of one of the two furnaces and sample holders mounted at the top of the basic unit. Sample holders, as illustrated in Fig. 8, are inserted into the furnace from the front of the unit. The use of two furnaces allows one to cool while the other is heating. Switching circuits are provided for additional external furnaces.

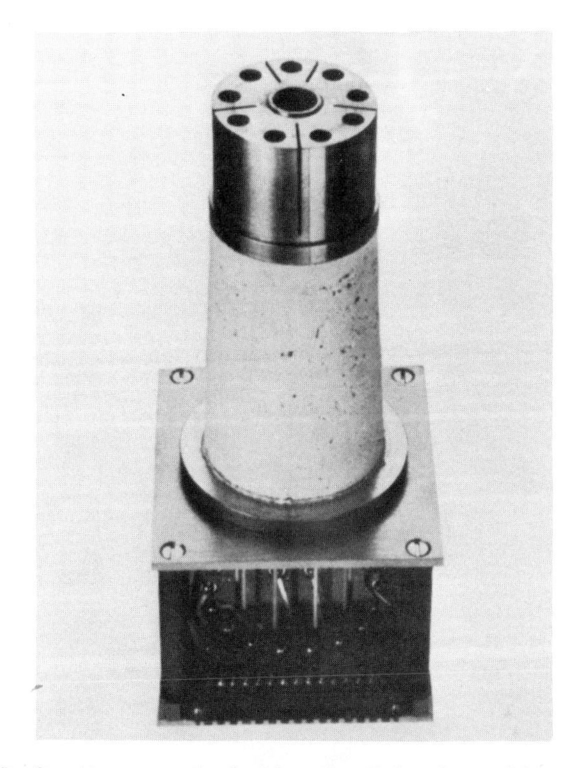

FIG. 8. Four sample holder for Deltatherm II system.

The various sample holder inserts are illustrated in Fig. 9. Each insert consists of a sample block, a cast ceramic supporting pillar, a printed circuit connecting board and connecting plug, and the apprpriate thermocouple or sample configuration. The standard insert (Fig. 8) has nine wells, so that up to four different sample-reference pairs may be studied simultaneously. Well size is 0.250 in. in diameter by 0.438 in. deep. All inserts are interchangeable and their maximum usable temperatures depend upon their material of construction.

Two furnaces are available. The F3001R furnace has a Kanthal heater element on an alumina core and is usable to 1250°C. The F3616 furnace has a platinum heating element on an

FIG. 9. Sample holder inserts for the Deltatherm II instrument.

alumina core and is usable to 1600°C. Both furnaces are of the vertical-tube, closed-top type, in which the sample holder insert is inserted from the bottom. Furnaces are interchangeable and may be flushed with any nonhazardous gas through an inlet fitting mounted on the furnace.

SPECIFICATIONS

Sensitivity: Standard amplifiers are 20 to 400 μV/in. in five steps. High gain amplifiers, which are optional, range from 1 to 400 μV/in. in nine steps.

Temperature range: Ambient temperature to 1250°C on D2000 Basic Unit; to 1600°C with the D2000-16 unit. A cryogenic accessory cell is available with a range of -200° to 700°C.

Programming rates: 2, 5, 10, and 20°C/min on D2200 Rate Control; optional Electronic Rate Reference and Chart Drive divide the D2200 heating rates into 100 equal increments.

Temperature indication: Sample block temperature is continuously displayed on a tape slide in 1° increments on the D2000 unit and 4° increments from 0° to 200°C and 2° increments above 200°C on the D2000-16 unit. Temperatures are indicated on the chart every 50° for the D2000 and every 100° on the D2000-16 unit.

Operating modes: Heat, cool, plus isothermal hold or automatic cutoff at present temperature. Temperatures may be preset every 50° with D2000 and every 100° with the D2000-16 unit. An optional Fine Temperature Limit Timer will divide the above preset increments into 60 parts.

Recorder: Four-channel, linear-response recorder uses electrosensitive paper.

Power requirements:

D2000: 100 to 130 V, 60 Hz, 1000 W.

D2000-16: 100 to 130 V, 60 Hz, 2500 W.

Dimensions and weight: 24 x 24 x 74 in. (61 x 61 x 181 cm); 365 lb. (166 kg).

Deltatherm III

The Deltatherm III consists of a table-top mounted control console with DTA furnace or TG plug-in modules. The recorder is built into the console for convenient display of the DTA or TG curve. The DTA cell consists of a quick-cooling furnace, a ceramic furnace support tube, and a variety of sample holders. A convenient feature is the break-in-two furnace which permits fast cooling by exposure to a small fan. Spare furnaces may also be used for fast rerun periods.

Three types of sample holders are provided: a) open interior blocks of Inconel or aluminum, b) well-type block of Inconel, and c) thermocups consisting of 1/8 in. or 3/16 in. diameter steel cups spot-welded on Chromel-Alumel wires. Microwells, which are ceramic sleeves enclosing the sample material around the bare wire thermocouple, are also used. The well-type sample block will accommodate thermocups, micro-well, bare wire thermocouples, or sheathed thermocouples.

The Deltatherm III control console and DTA module are shown in Fig. 10.

SPECIFICATIONS

Sample temperature indication: High-precision analog-digital display, ±3°C (dependent on thermocouple), reproducibility ±1/2°C.

Recorder output: Can be supplied for any recorder with input sensitivity up to 1 V/in.

Presentation: ΔT signal plotted continuously with accurate 100°C calibrations as pips, reproducible to ±1/4°C recorded in less than a second (event-marker output available for recorders so equipped).

Programmer: Full servofeedback electronic programmer.

Heating rates: 1.25°, 2.5°, 5°, 10°, and 20°C/min.

Rate linearity: ±3% from room ambient to 1200°C; ±2% from 200° to 800°C.

Repeatability: Better than ±1%.

Power requirements: 105 to 135 V ac, 50 or 60 cycle DTA.

Sample system: Standard wells, pan-type sample cups, or ceramic-sleeve holder easily interchangeable.

Thermocouples: Chromel-Alumel.

Sample size: 0.0001 to 0.5 g.

FIG. 10. Deltatherm III DTA system.

Sample atmosphere: Pressure to vacuum or controlled atmospheres. Atmosphere gas can be dynamically flowed through sample and collected if desired.

Furnace: Quick change, quick cooling, medium mass, Kanthal.

Temperature range: Ambient to 1200°C.

ΔT sensitivity: 0.025° to 10°C/in. in nine steps.

ΔT zero shift: 15 in. either way from center of chart.

Size: Control Console - 16 x 21 x 16 in.; DTA Module - 8 x 8 x 18 in.

Weight: Control Console - 24 lb; DTA Module - 8 lb.

Accessories:

Recorder: Any potentiometric strip-chart recorder may be used.

Thermogravimetric Analyzer: See Thermobalances, p. 163.

DU PONT INSTRUMENTS

The Du Pont Modular Thermal Analyzer System brings to the laboratory bench a rapid and convenient means of evaluating the thermal behavior of materials. The basic unit of the system is the versatile 990 Thermal Analyzer, into which are plugged a variety of modules and accessories. These include modules for differential thermal analysis, thermogravimetry, and thermomechanical analysis. The Du Pont instrumentation is an outgrowth of years of experience with units built "in-house" to meet the needs of Du Pont research and quality control laboratories.

990 Thermal Analyzer

The heart of the 990 Thermal Analyzer, as illustrated in Fig. 11, is the solid-state electronic temperature programmer-controller, which permits a wide range of temperature scanning conditions from -190° to 1600°C. Heating, cooling, isothermal, hold, and cyclic operations are selected by a "program mode" switch. The heating or cooling rates and the starting temperature of the program can be varied to suit individual needs.

The 990 Basic Unit contains the one- or two-pen X-Y recorder, which records on 11 x 17 in. chart paper. The Y axis is calibrated in units of the variable being recorded. Actual sample temperature is recorded on the X axis, which has suppression in units of 5°C/in.

FIG. 11. Du Pont 990 thermal analyzer, basic unit.

SPECIFICATIONS

Operating modes: Heat, cool, hold, isothermal, and cyclic.

Sample size: 0.1 to 5 g, depending on module.

Heating rates: 0.5 to 100°C/min, continuously variable.

Heating rate accuracy: ±5% or 0.1°C/min, whichever is greater.

Heating rate stability: Less than 5% change in rate for a 20 V change in line voltage.

Drift in isothermal position: Less than 1°C/hr.

Drift in hold position: Less than 3°C/hr.

Size and weight: 22 1/4 x 20 3/4 x 10 3/8 in.; 50 lb.

Utilities: 115 V, 50 or 60 Hz.

Y axis scale: 0.004, 0.008, 0.02, 0.04, 0.08, 0.2, and 0.4 mV/in.

X axis scale: 0.4, 0.8, 2.4, and 8 mV/in.

Chart speed for time base: 0.05 to 10 in./min.

All DTA and DSC modules plug in to the Cell Base, which in turn is connected to the 990 Basic Unit. The Cell Base, as illustrated in Fig. 12, contains the electronic cold junction compensator; the variable, resettable baseline slope control; and the gas atmosphere and cooling inlets. The Cell Base accepts the cells described in Table 2.

TABLE 2

DTA and DSC Cells for 990 Cell Base

	T-Range, °C	Sensitivity
Standard DSC	-190° to 600°	0.05 mcal/sec/in.
Pressure DSC	-190° to 600°	0.05 mcal/sec/in.
Standard DTA	-190° to 500°	0.05°C/in.
Intermediate DTA	Room to 850°	0.05°C/in.
1200°C DTA	Room to 1200°	0.05°C/in.
1600°C DTA	Room to 1600°	0.05°C/in.

DTA Cells

The plug-in module type DTA cells are illustrated in Fig. 13.

FIG. 12. Du Pont cell base.

The Standard DTA Cell (500°C) is a general-purpose, block-type cell. Samples are placed in disposable glass tubes (2 and 4 mm in diameter) into which are inserted disposable Chromel-Alumel thermocouples. A variety of accessories are available for the cell, including Visual Cells for observation of the sample during analysis, Cooling Jacket, Quick-Cool Accessory, and Pt, Pt-13% Rh thermocouples. The specifications are given in Table 3.

The Intermediate Temperature DTA Cell (850°C) is similar to the Standard Cell, except that it uses quartz tubes in place of glass and attains a higher temperature at a small sacrifice in quick-cooling capacity.

The High Temperature DTA Cell plug-in module permits measurements to 1200° or 1600°C. The low thermal mass of the

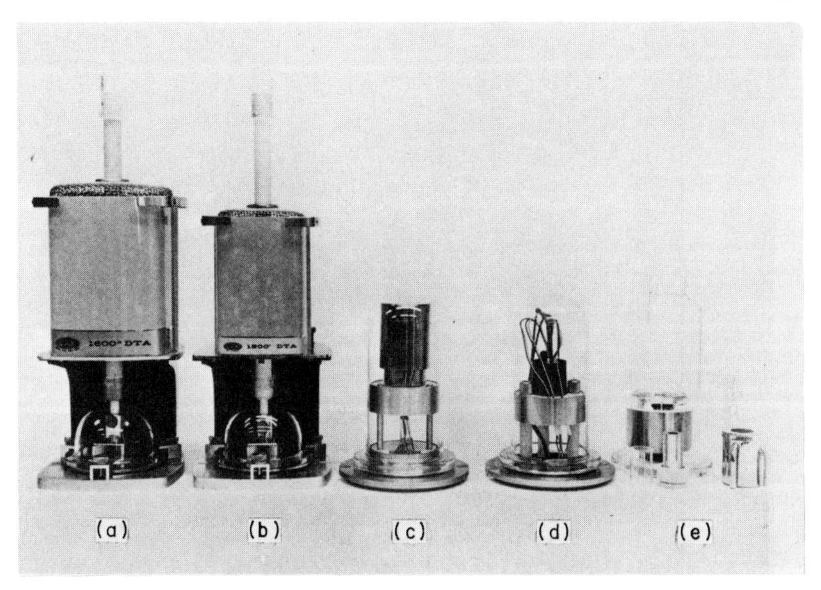

FIG. 13. Du Pont DTA cells: (a) high temperature cell (1600°C); (b) high temperature cell (1200°C); (c) intermediate temperature cell (850°C); (d) standard DTA cell (500°C); (e) cooling jacket.

furnace and its modular design permit rapid cooling. Alternatively, a hot furnace may be quickly replaced by a cool one and a new sample inserted, all within ten minutes. The 1200°C furnace is Kanthal wire-wound, while the 1600°C furnace uses Pt - 5% Rh alloy wire. The furnace permits intimate exchange between the sample and the controlled atmosphere. Connections on the basic unit allow the cell to be evacuated or purged with inert gases. Samples are retained in removable platinum sample and reference cups in two sizes.

DSC Cell

The Differential Scanning Calorimeter (DSC) Cell, as shown in Fig. 14, has a pan-type sampling system with provision for

TABLE 3

Specifications for Du Pont DTA Cells

Parameter	Standard cell	Intermediate cell	High-temperature cell
Temperature range	-100° to 500°C Accessories extend low range to -190°C	25° to 850°C	25° to 1200° or 1600°C
ΔT sensitivity	0.004 to 0.4 mV/in. 0.1 to 10.0°C/in.		0.3 to 34°C/in.
Sample size	0.1 to 80 mg		Up to 75 mm^3
Sample tubes	Glass, 2 and 4 mm diameter; quartz, 2 and 4 mm; Pt cups, 3 and 5 mm.		
Thermocouples	Chromel-Alumel		Pt, Pt-13%Rh
Atmosphere	Static or flow; air, inert, or reactive gas		
Pressure	1 atm to 1.0 Torr		1 atm to 2 Torr
Temperature precision	±0.2°C		±2.0°C
Program cooling rate	0.5 to 20°C/min	None	To 15°C/min

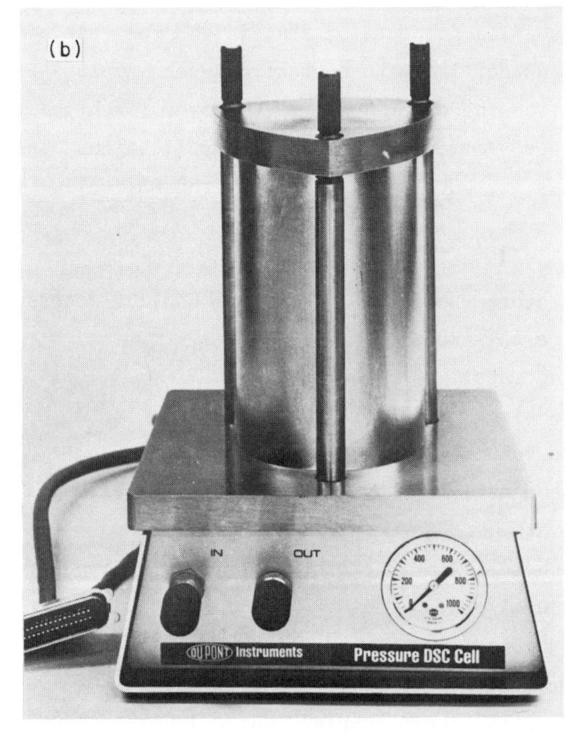

FIG. 14. Du Pont differential scanning calorimetry cell: (a) standard cell; (b) high pressure cell.

hermetic sealing of the samples. With the DSC cell, temperature differences between sample and reference are recorded on the Y axis of the basic unit recorder, and sample or reference temperature is recorded on the X axis. The pen deflection in the Y direction is proportional to the difference in the quantity of heat flowing to the sample and to the reference per unit time. Thus, the area under the peak is directly related to the amount of energy (calories) involved in the transition under examination. This relationship is found by reference to a calibration curve determined with samples of known enthalpy (ΔH).

In general, the DSC cell is recommended for most accurate quantitative data on heats of transition, for ease of sample handling, and for intimate contact between sample and environment. DTA, on the other hand, is employed for highest temperature accuracy, for certain corrosive applications, and where temperatures are outside the range of the DSC cell.

SPECIFICATIONS

Temperature range: 25° to 600°C; Cooling accessories extend range to -120°C.

Cooling rate: 10°C/min to -110°C.

Sample size: 0.5 to 100 mg.

Sample volume: 0.05 ml.

Sample pans: Aluminum; open or hermetically sealed to 3 internal atm.

Atmosphere: Atmospheric to 2 Torr; preheated dynamic gas purge to 100 ml/min.

Cell volume: 2 ml.

Temperature repeatability: ±1°C.

Differential thermocouple: Chromel-Constantan.

Sample thermocouple: Chromel-Alumel.

Calorimetric Sensitivity: 0.2 mcal/sec.

Calorimetric precision: ±1%, based on fusion of metal samples.

The Pressure DSC Cell, Fig. 14(b), provides calorimetric data and transition temperature data from 0.01 Torr to 1000 psig. Specifications are similar to the DSC cell previously discussed except that the pressure may be varied from 0.01 Torr to 1000 psig, with dynamic gas purge.

EBERBACH CORPORATION

Portable Differential Thermal Analysis Apparatus

This unit is primarily a portable field instrument designed to enable the operator to obtain qualitative and rough quantitative data on certain mineralogical samples. It is not intended to replace laboratory instruments but may be used to supplement them, as a screening instrument for rapidly checking samples in order to select the more promising ones for detailed treatment later on. The chief advantage of the unit lies in the fact that a determination can be made in about 20 minutes. The complete apparatus is illustrated in Fig. 15.

The test procedure consists of placing the sample and the inert material into the separate cells or a crucible, and then manually plotting the difference in temperature between them (as the ordinate) against the temperature of the crucible (as the abscissa). Difference in temperature, ΔT, is read from a galvanometer which is connected to the thermocouples in the sample and reference wells. The crucible temperature is read from the pyrometer, which is connected to a thermocouple in the third cell of the crucible.

The apparatus has three of the crucible assemblies with three cells each. Rapid heating is accomplished by a small cylindrical furnace, which, after being preheated, is slipped over the crucible to be used for the determination. A three-position galvanometer sensitivity switch provides low, medium, and high. A four-position selector switch provides a position for verifying or adjusting the mechanical zero of the galva-

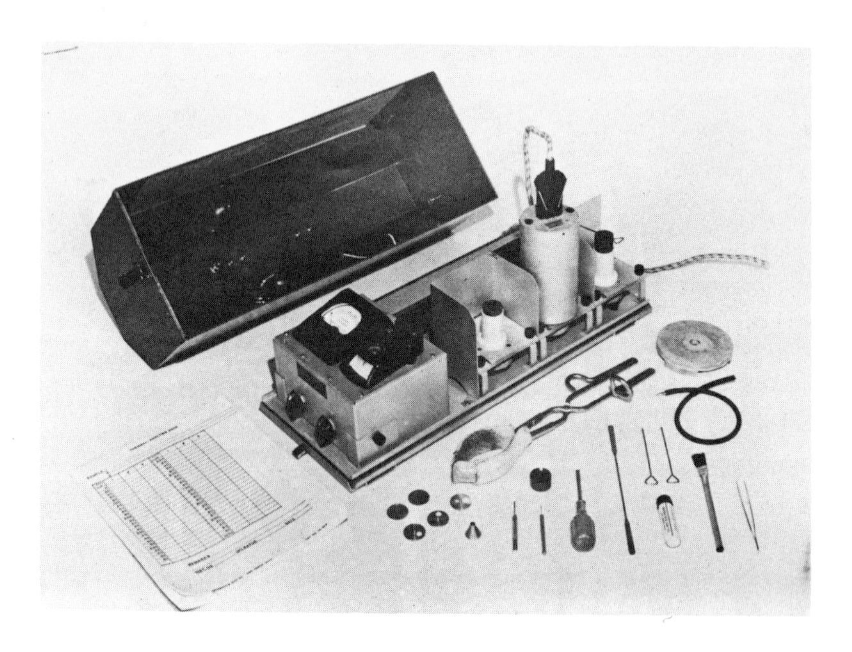

FIG. 15. Eberbach portable DTA apparatus.

nometer and three positions by which the temperature-indicating thermocouple in the crucible being heated can be connected with the pyrometer. Maximum temperature of the furnace is 900°C.

The portable unit is enclosed in a metal case measuring 21 x 8.5 x 9 in. The unit operates on 115 V ac or dc.

FISHER SCIENTIFIC

Series 200 Differential Thermalyzer

This is a low-cost differential thermal analysis system designed for educational, research, and industrial quality-control laboratories. The system, as illustrated in Figs. 16 and 17, consists of three separate units: a) sample holder, which has provision for two sets of reference and sample thermocouples and crucibles; b) furnace, capable of use up to 1200°C; and c) programmer. The user is required to supply a recorder with a sensitivity of at least 1 mV full-scale. A microvolt dc amplifier is not supplied with the system; hence, a larger sample size, 50 to 200 mg, is required.

The sample holder has an insulated handle and indexing pins which automatically center the Inconel block in the top of the furnace. Wells bored in the Inconel block contain the control thermocouple and two pairs of quartz crucibles for the samples and reference materials. If two samples are to be run simultaneously in the system, two recorders must be employed, one for each sample-reference combination. A removable rack, which can be mounted on either side of the furnace, holds the sample assembly while the block is cooling or being loaded.

The furnace programmer contains a proportional controller that increases the power to the furnace at a rate that will give the desired temperature increase -- 5°, 10°, or 25°C/min. The programmer uses a control thermocouple feedback system. A dial, rotating in synchronization with the programming potentiometer, indicates the furnace temperature to within ±2%. The

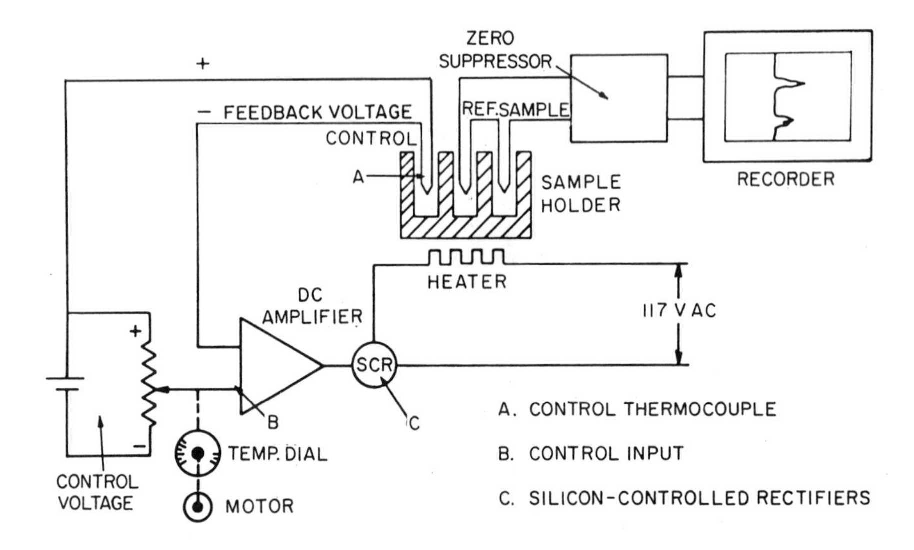

FIG. 16. Schematic diagram of Fisher differential thermalyzer.

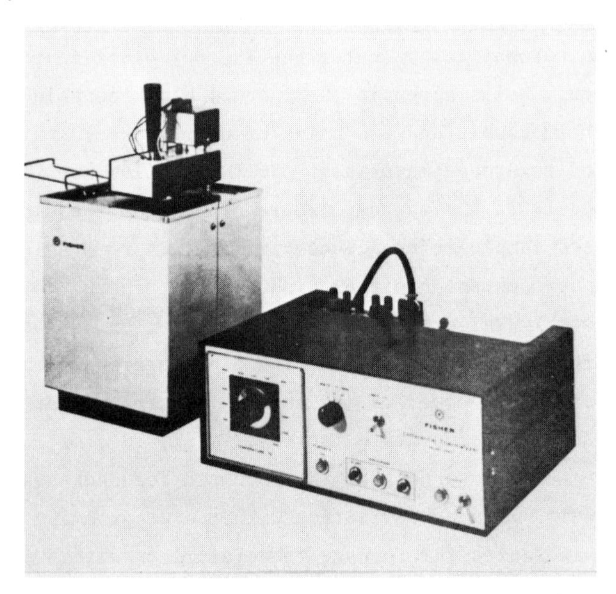

FIG. 17. Photograph of Fisher differential thermalyzer.

exact furnace temperature must be obtained from a temperature-time calibration curve, previously obtained. A selector switch in the unit determines the maximum furnace temperature, either 500° or 1200°C.

Three different types of thermocouples are supplied with the system, enabling the user to select the thermocouple best suited to the desired temperature range. The iron-Constantan thermocouples may be used up to 350°C; Chromel-Alumel thermocouples, up to 850°C; and Platinel thermocouples, up to 1200°C. Zero suppressors for one (or two) recorders are built into the programmer and are used to center the recorder tracing on the chart.

The complete system includes the programmer, one furnace, a sample-holder assembly, a dozen quartz crucibles, two sets of three different sample thermocouples, one Platinel calibration thermocouple, a pound of alumina powder, a pound of potassium sulfate as a test material, and a stainless steel micro spatula.

SPECIFICATIONS

Temperature range: Ambient to 1200°C.

Programmer ranges: Ambient to 500°C; ambient to 1200°C.

Programming rates: 5°C/min ± 10%; 10°C/min ± 10%; 25°C/min ± 10%.

ΔT sensitivity: 0.25°C with 1 mV recorder.

Sample size: 50 to 200 mg.

Furnace atmosphere: Air.

Voltage requirements: 115 V, 60 cycles; 230 V, 50 cycles.

Power requirements: 500 W.

Weight:

- Sample holder: 1 1/2 lb.
- Furnace: 12 7/8 lb.
- Programmer: 10 1/2 lb.

Size:

Furnace: 8 x 8 x 10 in.

Programmer: 13 3/4 x 10 (FIB) x 5 1/4 in.

Accessories or complementary equipment: 1) Temperature programmer; 2) Thermobalance accessory (see p. 165).

Series 200A Differential Thermalyzer

This is a DTA system with thermal specifications similar to the Series 200 unit. It is a complete DTA system and includes a programmer and two-pen recorder. The programmer is described in the Series 300 QDTA system. The Series 200A unit permits operation in controlled atmospheres, over a temperature range from -100° to 1200°C. An electronic temperature compensator enables temperatures to be read with an accuracy of ±0.5°C.

Series 300 QDTA System

The Series 300 Quantitative Differential Thermal Analysis (QDTA) system is a modular DTA system with the stability and reproducibility necessary for accurate calorimeter measurements. A special sample module and amplifier give the QDTA system the sensitivity to measure glass transition energies, small changes in specific heat, and other small thermal effects. A schematic diagram of the QDTA system is shown in Fig. 18.

The Series 300 sample module has a pan-type cell, in which the temperature-sensing junctions are platinum disks welded to Platinel II thermocouples. These sensors are permanently aligned and are outside of the sample, so that changes in the thermal conductivities of the materials do not introduce significant errors. Disposable aluminum sample pans are used, which can be hermetically sealed so that volatile samples may be studied.

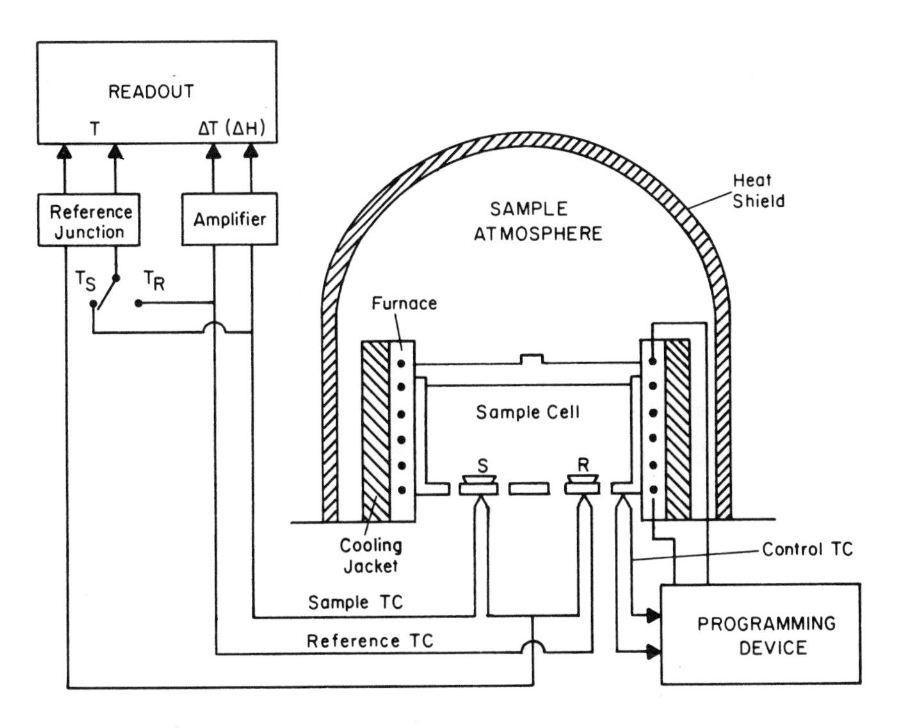

FIG. 18. Schematic diagram of Fisher Series 300 QDTA system.

The furnace has a long uniform-temperature zone that encloses the entire sample chamber. A cooling jacket makes it possible to program the system at a reverse rate: air cooling gives a reverse program rate up to -10°C/min, while tap water will give rates up to -25°C/min. Liquid nitrogen cooling will enable temperatures to -150°C to be attained.

Built-in connectors permit the circulation of inert or reactive gases through the sample chamber. Pressures up to 1 atm and flow-rates up to 200 ml/min may be employed. By placing a bell jar over the sample chamber, the system can be operated under pressures as low as 1 Torr.

SPECIFICATIONS

Temperature range: -150° to 600°C.

Temperature program modes: Linear increase from isothermal: a) hold at limit, and b) off at limit; Linear decrease from isothermal: a) hold at limit, and b) off at limit; isothermal.

Program limits: Upper and lower limits continuously variable.

Program rates: 0.5°, 1°, 2°, 2.5°, 5°, 10°, 20°, and 25°C/min. Continuously variable: 0.1° to 150°C/min.

Accuracy: ±2%.

Reproducibility: ±0.5%.

Linearity: ±1% of full scale.

Isothermal control: ±0.25% or ±3°C.

Temperature stability (ambient: 15°-40°C): ±0.03%/1°C.

Line voltage stability (120 V ±10%): ±1°C.

Sample size: 0.01 mg minimum; 100 mg maximum.

Sample atmosphere: Vacuum or noncorrosive gas purge.

Sensitivity (ΔT): 0.1°C/in; 0.15 cal/sec/in.

T ranges (approximate): Full-scale ranges with Platinel I or Chromel-Alumel thermocouple: 50 mV - 1300°C; 40 mV - 1000°C; 20 mV - 500°C; 10 mV - 250°C; 5 mV - 125°C.

Range suppression: -1 to +5 ranges in six steps.

Precision (1/40 in. on chart): ±0.2°C (5 mV range).

Readability (1/20 in. on chart): ±0.5°C (-150°/1200°C).

Presentation [10-in. dual-channel readout (Y and Y' vs. time)]: Simultaneous presentation of ΔH and ΔT vs time.

Power requirement: 7 A on 115 V ac, 60 Hz; 3.5 A on 230 V ac, 50 Hz.

Overall dimensions (maximum, complete system): 36 x 15 x 13 in.

Shipping weight: 84 lb.

HARROP LABORATORIES

Model DTA-T1

The Model DTA-T1 DTA instrument is illustrated in Fig. 19.

This instrument is designed for DTA applications requiring the highest possible accuracies, combined with maximum versatility. It functions equally well for rapid routine studies or for more in-depth research applications. The instrument is self contained except for an externally mounted recorder. Five furnace options are available: (a) one 1550°C and one 1200°C furnace; (b) two 1550°C furnaces; (c) two 1200°C furnaces; (d) one 1550°C furnace; and (e) one 1200°C furnace. Each furnace is stainless-steel jacketed and is water cooled for cool external operation. A Kanthal heater element is used for the 1200°C furnace and a platinum element for the 1550°C furnace.

A nickel sample block is supplied with furnaces for operation to 1200°C while an alumina block is used for the higher temperature system. Other block and sample cup materials are available. Matched thermocouples are used combined with guide rods to align the sample block and furnace; use of both of these is said to give better reproducibility of the curves and a minimum of base-line drift.

SPECIFICATIONS

Temperature range: 1550°C in platimum element furnace, 1200°C in Kanthal element furnace.

Furnace jackets: Water-cooled stainless steel and aluminum end plates.

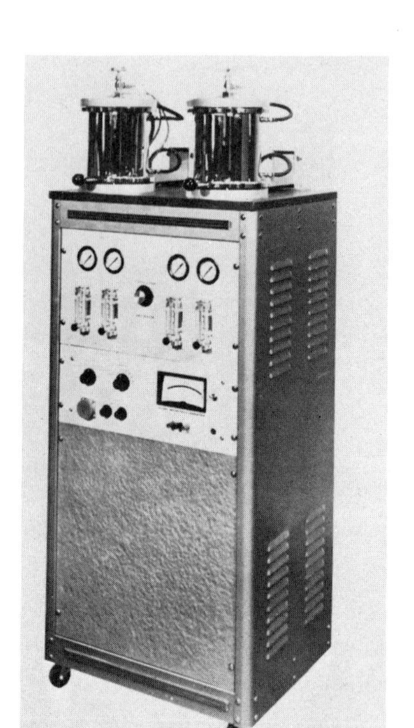

FIG. 19. Harrop Model DTA-A1 DTA instrument.

Sample containers: Nickel, alumina, and beryllia blocks; alumina and platinum cups; platinum sleeves.

Thermocouples (differential): Pt/Pt 10% Rh matched to within 3 μV.

Thermocouples (furnace control): Pt/Pt 10% Rh.

Sample capacity: Micro to 1.25 cm^3.

Environment: Air, vacuum, and noncorrosive atmospheres.

Pressure: 10^{-3} Torr to 30 psia.

Gas-flow controls: 0 to 5 LPM.

Gas pressure controls: 0 to 15 psi.

Amplifier sensitivity: 0.3 μV/in. on recorder (0.03°C/in.)

Amplifier stability: 0.5 μV/day.

Power: 115 V 60 Hz.

Coolant: Tap water at 1 LPM below 60°F.

HUNGARIAN OPTICAL WORKS

Derivatograph

The Derivatograph is a multifunction thermal analysis system which can record on a single chart the TG, DTG, DTA, and T curves of a sample. By means of a simple attachment, the TD (thermal dilation) and DTD curves of the sample can also be recorded. Evolved gas analysis may also be carried out under certain conditions (thermogas analysis).

The instrument consists of the following components: a) analytical balance with semiautomatic weight loading device galvanometers, b) two furnaces, c) sample and reference material crucibles, d) furnace temperature controller, e) voltage regulator, and f) galvanometric light beam-photographic paper recorder. These components are illustrated in Figs. 20 and 21.

The balance is an air-damped analytical type with a basic sensitivity of 20 ± 0.2 mg full-scale deflection. The working range of the instrument is 10 mg to 10 g. The sensitivity of the balance can be changed by means of an automatic weight loading device and can be varied in steps of 20, 50, 100, 200, 500, 1000, and 2000 mg f.s.d. The accuracy (except for the 20 and 50 mg f.s.d.) is within the limits of ±0.2%.

The derivative of the TG curve is obtained by a simple device consisting of a magnet and an induction coil. The permanent bar-magnet is suspended on one arm of the beam of the balance with both of its poles surrounded by two coils with a large number of turns. No voltage is induced into the coils until the balance beam is deflected by a mass change of the

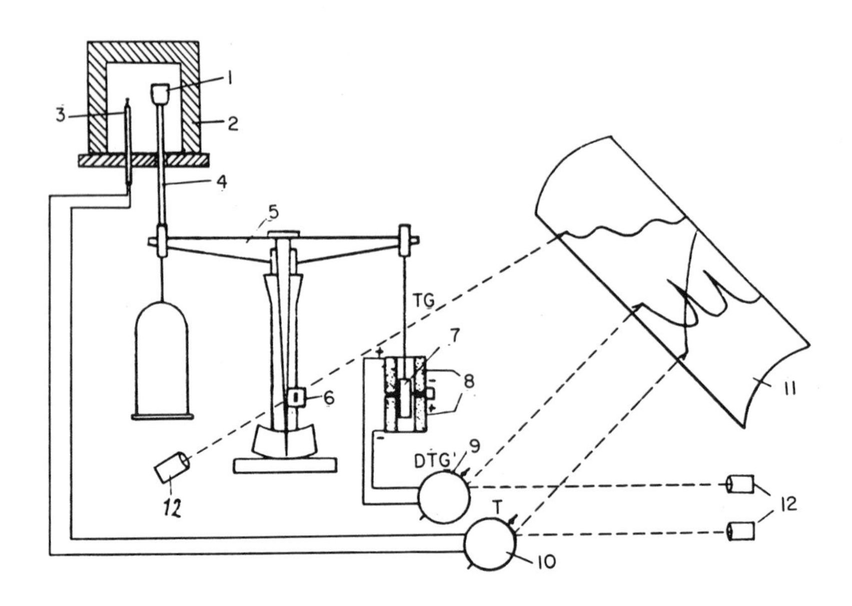

FIG. 20. Derivatograph, schematic diagram. 1 - Crucible for the sample, 2 - electric furnace, 3 - thermocouple, 4 - porcelain rod, 5 - balance, 6 - optical slit, 7 - permanent magnet, 8 - coils, 9, 10 - galvanometers, 11 - photographic paper, 12 - lamps.

sample. When such a mass change occurs, the magnet moves and a voltage proportional to the rate of change is induced in the coil. The induced coil voltage is then measured by one of the light beam galvanometers and recorded on the chart.

SPECIFICATIONS

Galvanometers: Sensitivity of the T galvanometer is 0-150°, 300°, 600°, or 1200°C, f.s.d. Accuracy is ±0.5%. Sensitivity of the DTA and DTG galvanometers is 1/1, 1/1.5, 1/2, 1/3, 1/5, 1/10, 1/15, 1/20, 1/30, 1/50, 1/100, and 1/200.

Sample holder crucibles: Standard sample holders are three pairs of platinum crucibles of different sizes to accommodate

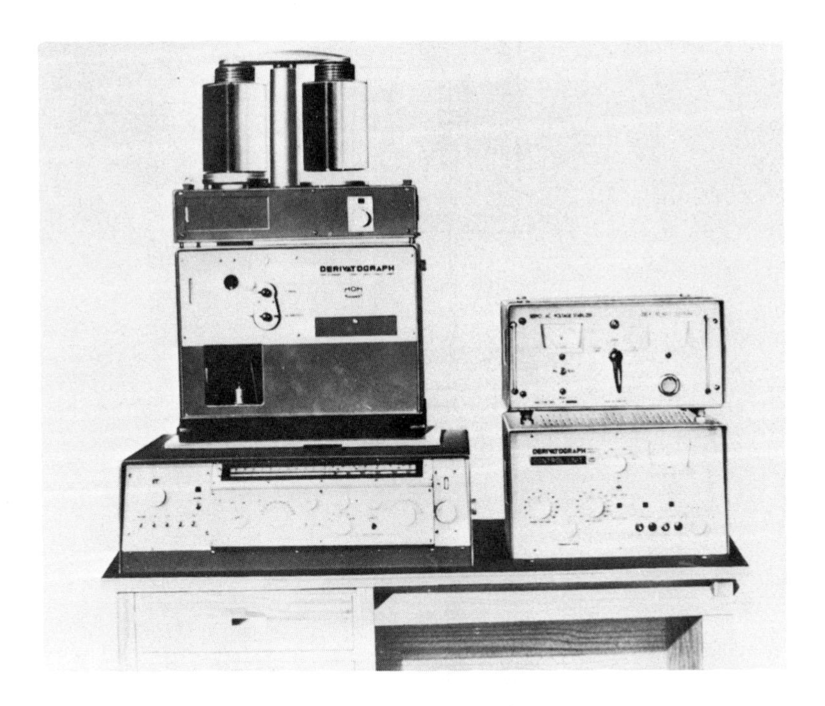

FIG. 21. Derivatograph, photograph.

sample masses from 50 to 2000 mg. Optional sample holders are available.

Recorder and photorecording drum: The drum is protected from room light and can be removed from the instrument for loading in a dark room. Drum can be rotated at speeds of 25, 50, 100, and 200 min/revolution.

Furnace: Twin electric furnaces are supplied which can be heated to a maximum temperature of 1100°C. Furnace atmospheres may be N_2, CO_2, Ar, O_2, etc., at atmospheric pressure.

Furnace temperature programmer: Controlled by a disk-type program-control unit which permits heating rates of 0.5°, 1°, 2.5°, 5°, 10°, and 20°C/min. Constant temperature of the furnace may be obtained by use of the temperature regulator.

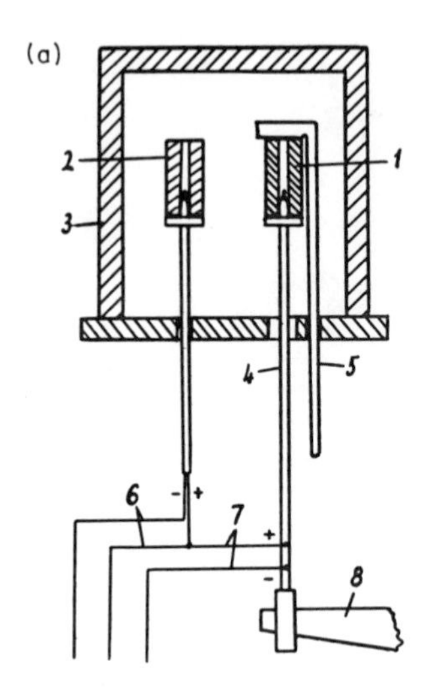

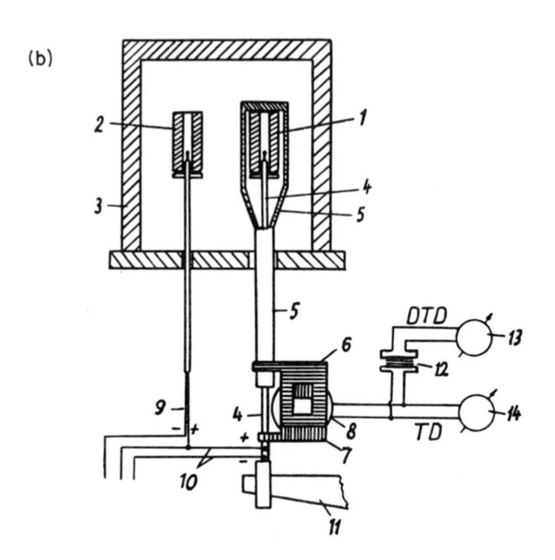

FIG. 22. Dilation devices for derivatograph. (a) TD and DTD curves only; 1 - compressed test piece, 2 - compressed reference substance, 3 - electric furnace, 4 - silica tube, 5 - silica rod, 6,7 - thermocouples, 8 - balance. (b) Simultaneous DTA, T, TG, DTG, TD, and DTD curves; 1 - compressed test piece, 2 - compressed reference substance, 3 - electric furnace, 4 - silica tube, 5 - silica stirrup ending in a tube, 6,7 - diaphragms, 8 - light cell, 9,10 - thermocouples, 11 - balance, 12 - transformer, 13,14 - galvanometers.

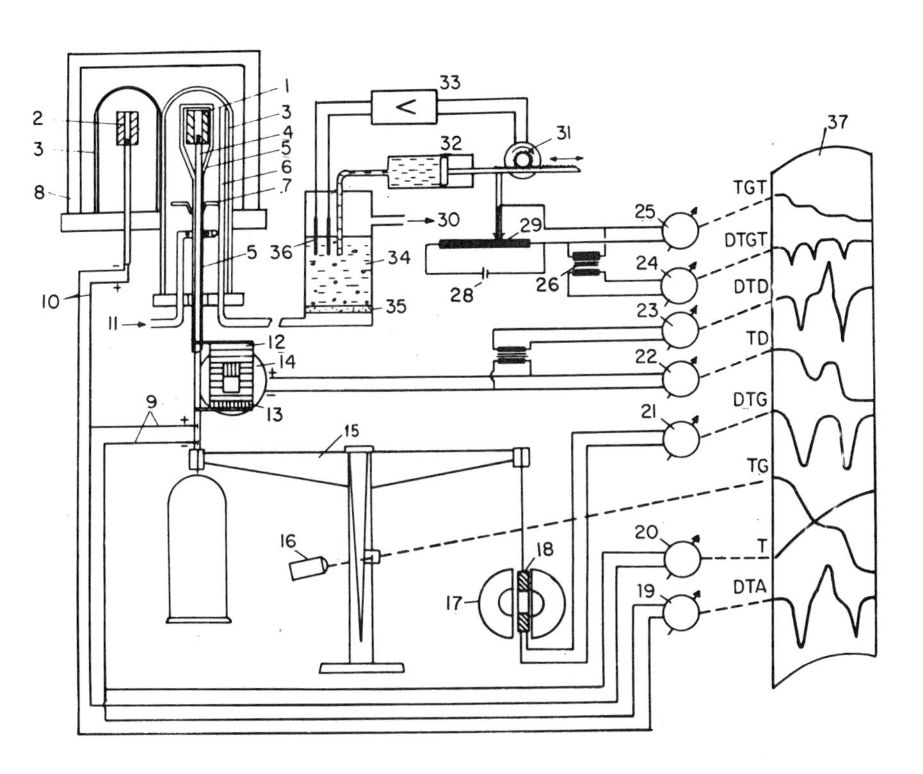

FIG. 23. Derivatograph thermo-gas analysis. 1 - sample, 2 - inert substance, 3 - silica bells, 4 - silica tube supporting the sample, 5 - silica tube resting on the sample, 6 - silica tube for gas extraction, 7 - silica diaphragm, 8 - electric furnace, 9-10 - thermocouples, 11 - silica tube for carrier gas introduction, 12-13 - diaphragms, 14 - photoelectric cell, 15 - balance, 16 - lamp, 17 - permanent magnet, 18 - coil, 19-25 - galvanometers, 26-27 - deriving transformers, 28 - electrical source, 29 - potentiometer, 30 - vacuum pump, 31 - servomotor, 32 - automatic burette, 33 - amplifying and controlling device, 34 - absorber, 35 - glass frit, 36 - electrodes, 37 - photographic paper.

The thermal dilation curves of a sample may be obtained by use of the simple device illustrated in Fig. 22. The sample is compressed into a cylindrical shape and rests on a quartz tube attached to one arm of the balance. It is brought into contact with the fixed quartz rod by loading the other arm of the balance with several hundred milligrams of excess weight. If the length of the sample changes during heating, the balance will indicate a deflection. The light signal from the balance therefore records the thermal dilation curve, and the galvanometer of the derivative device records the derivative dilation curve (DTD). The instrument will also simultaneously record the DTA and T curves. The quartz rod can then be removed and the procedure repeated with another sample of similar dimensions, so that the TG, DTG, DTA, and T curves can be recorded. If all six curves of the same sample are required from a single run, the attachment shown in Fig. 22 may be employed.

For thermogas analysis, the modification illustrated in Fig. 23 may be employed. The sample is surrounded by a quartz tube and evolved gases are flushed to the absorber by an inert carrier gas. Volatile components (e.g., sulfur trioxide) that may condense in the cooler parts of the system are washed into the absorber by a slow stream of water. The evolved compounds in the absorber solution are then continuously titrated, either manually or potentiometrically. It is possible to monitor the evolution of SO_2, SO_3, CO_2, HCl, HI, Br_2, I_2, and other gases.

LINSEIS MESSGERATE GMBH

There are a large number of DTA instruments and/or sample-holder/furnace combinations available from Linseis. The various combinations of modular components are very versatile and cover a wide range of DTA applications in the temperature range -200° to 1500°C. Only two of the systems will be discussed here, the L62 system, which is a general purpose apparatus and can be used for almost any type of DTA study up to a maximum temperature of 1500°C, and the L63 system, termed a dynamic microcalorimeter (DMC), which covers the range -120° to 450°C and has a high ΔT sensitivity (0.0057°C/cm) and resolution (0.002°C). The latter is used primarily for heats of transition or reaction measurements and is said to have a ΔH reproducibility of ±1% (based on the fusion of pure metals).

The L62 System

The L62 DTA system employs a low thermal capacity measuring system with interchangeable sensors suitable for solids, powders, and liquids. Two furnaces are available, the L62/800 for the temperature range -150° to 300°C, and the L62/240 for the temperature range 20° to 1550°C. Both furnaces are shown in Fig. 24. Several sample holders are available, ranging from the platinum sleeve type to alumina crucibles. Sample volumes range from 0.016 to 0.170 cm^3.

Various sample holders are available, such as are illustrated in Fig. 25. The holders are interchangeable and are attached to the system by means of the carrier, as illustrated.

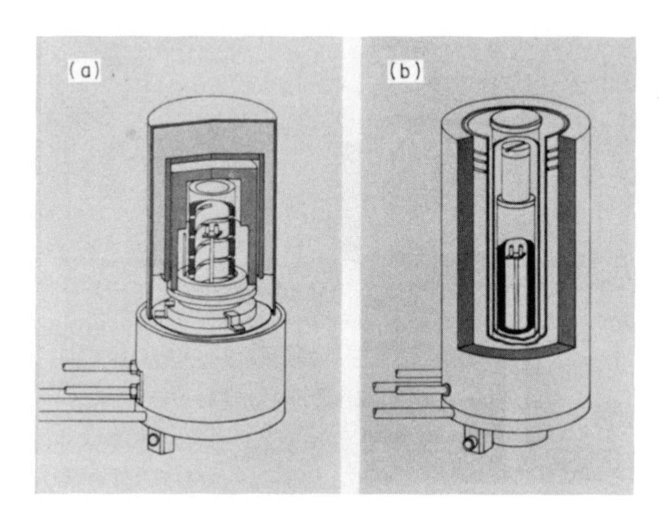

FIG. 24. Linseis L62 DTA system. (a) The L62/240 furnace; (b) the L62/800 furnace.

They are constructed of platinum, alumina and other materials and can be used with Pt-Pt/Rh, Chromel-Alumel, or Cu-constantan thermocouples.

SPECIFICATIONS

Temperature ranges: -150° to 300°C; 20° to 1500°C.

Atmosphere: Air or inert gases at atmospheric to 1×10^{-3} Torr.

Maximum ΔT sensitivity: 2 V/cm or 1 mcal if used in DSC mode.

Reproducibility: ±1% (for ΔH measurements).

Recorder: 16 ranges from 1 mV to 200 V; for ΔT measurements, 0.05, 0.1, 0.2, 0.5, 1, 2, 5, and 10 mV.

Controller: High temperature, Pt-Pt/Rh, 20° to 1600°C; Low temperature, Pt 100, -200° to 500°C; Heating rates, 0.1 to 50°C/min.

Output voltage: 35 V/45 V at 1.5 kVA.

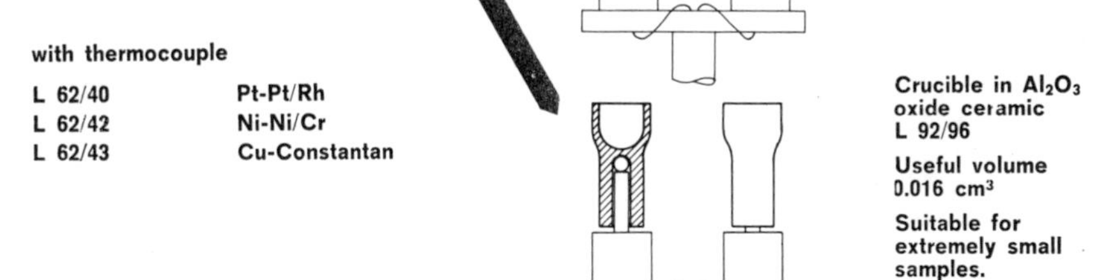
Interchangeable sample holders
Carrier
Pt sleeves
L 62/70
Useful volume
0.170 cm³
Suitable for powders which do not react with platinum.
Pt crucible
L 62/94
Useful volume
0.120 cm³
Suitable also for plastics.
This form can also be supplied in quartz (L 62/98)
Crucible in Al₂O₃ oxide ceramic
L 62/97
Useful volume
0.090 cm³
Suitable for metals and materials reacting with platinum.
with thermocouple
L 62/40 Pt-Pt/Rh
L 62/42 Ni-Ni/Cr
L 62/43 Cu-Constantan
Crucible in Al₂O₃ oxide ceramic
L 92/96
Useful volume
0.016 cm³
Suitable for extremely small samples.

FIG. 25. Linseis L62 DTA system sample holders and carrier.

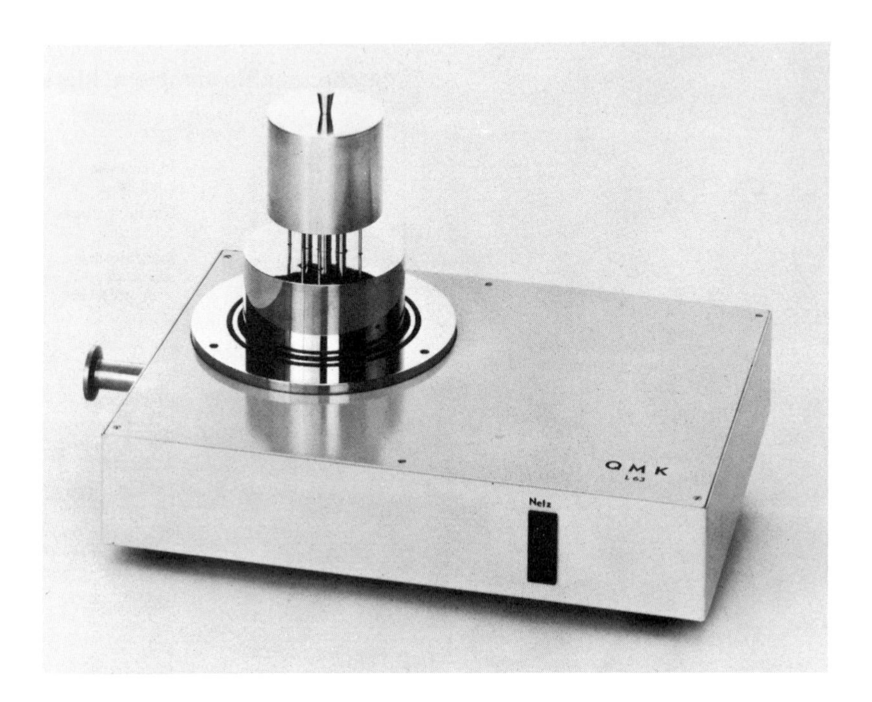

FIG. 26. Linseis L63 dynamic micro calorimeter (DMC); sample holder only shown.

The L63 System

As mentioned earlier, this apparatus is intended for quantitative DTA and specific heat measurements in the temperature range from -120° to 450°C. The sample holder, which supplements the L62 System, is shown in Fig. 26. For calorimetric measurements, the sample holder is calibrated using metals with known heats of fusion.

SPECIFICATIONS

Sample size: 0.1-100 mg.

T range: -120° to 450°C.

ΔT range: 0.01, 0.02, 0.05, 0.1, 0.2, and 0.5 mV/250 mm.

ΔT sensitivity (max.): 0.0057°C/cm.

ΔT resolution (max.): 0.002°C max.

Noise of ΔT amplifier: 0.0007°C.

ΔH resolution (max.): 1 mcal.

dH/dt (max.): 0.04 mcal/sec/cm.

Reproducibility of ΔH determination (for fused metals): ±1%.

Built-in electronic reference junction: 50°C ± 0.2°C.

Stability of reference junction: ±0.05°C.

T resolution (Ni-Ni/Cr thermocouple): 0.5° C.

Absolute accuracy: to DIN 43710 for Ni-Ni/Cr thermocouple (better than ±1°C after calibration).

Gas flow: up to 100 ml/min.

Vacuum: 0.1 Torr.

METTLER INSTRUMENT CORPORATION

The Mettler Thermoanalyzer is the most elaborate, versatile, and expensive of all the TGA-DTA systems commercially available today. It is truly a universal research instrument, which can simultaneously plot, on a single chart, the sample mass change as a function of time or temperature, the derivative of the mass change, the DTA curve, the pressure, and the flow velocity of the gas stream.

The thermoanalyzer is illustrated in Figs. 27 and 28. In Fig. 27, the balance and furnace assembly are shown mounted in the cabinet at the left while the control cabinet is on the right. The balance cabinet contains the vacuum system and all controls for the gas flow into the sample chamber. A complete water system is built in for cooling of the diffusion pumps and protection of the balance from the furnace temperature changes. The control cabinet contains the multipoint recorder, the temperature programmer, and measurement and control circuits for all of the balance functions. The electronic circuitry is all solid state.

The balance (as shown in Fig. 28) consists of an aluminum beam substitution type balance using sapphire knife edges and planes. The mass measurement system consists of an electrical coil, attached to the end of the beam, which moves vertically around a permanent magnet attached to the end of the beam, which moves vertically around a permanent magnet attached to the balance housing. Changes in sample mass cause a beam deflection, which moves a light shutter interrupting a light beam display

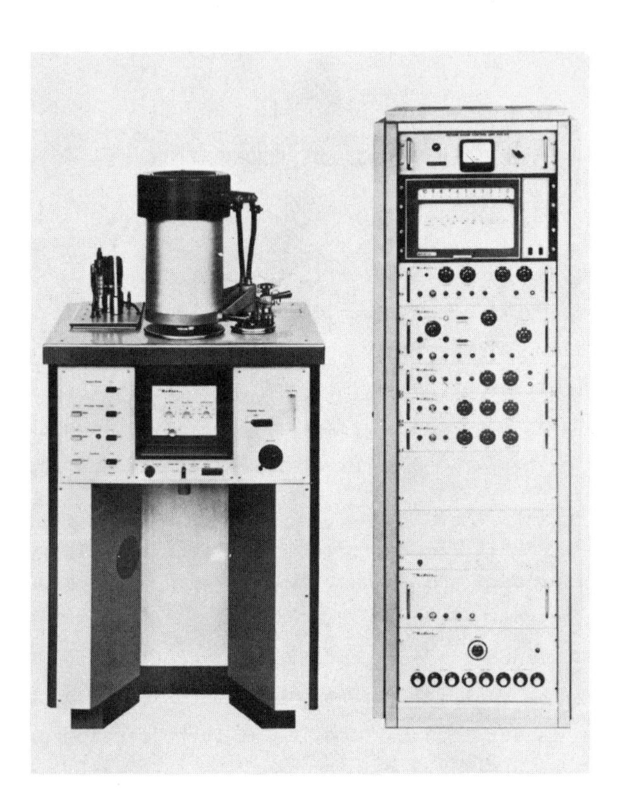

FIG. 27. Mettler recording vacuum thermoanalyzer (photograph).

on two photodiodes. The unbalance in photodiode current is amplified and fed back to the coil on the balance beam as a restoring force. Total balance capacity is 16 g, weighing above the beam, or 42 g below the beam. Vacuum-sealed control rods operate a weight set of 0 to 15.9 g, which is used as a tare for sample holder and support rod compensation.

The electrical weight indication has a dual weighing range with three different sensitivities as standard (a fourth is optionally available). Two consecutive sensitivities, in the ratio of 1:10, are always recorded. One range is 0-1000 mg,

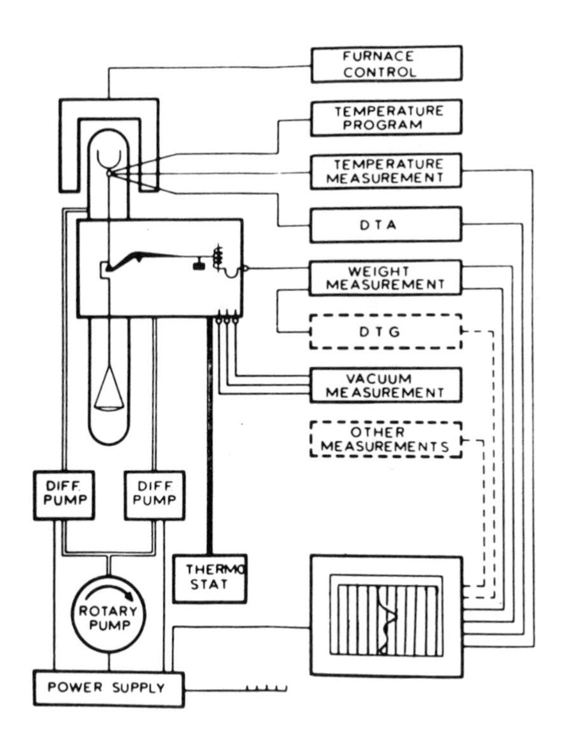

FIG. 28. Schematic diagram of the Mettler recording vacuum thermoanalyzer.

recorded as 100 mg/in., on the 10-in. chart. The second curve is the same 1000 mg as 100 in. full scale, or 10 mg/in. This is done by recording the curve on the chart ten times. A more sensitive weight range of 0-100 mg is recorded in an identical manner as above. An additional high sensitivity weight range of 0-10 mg, recorded in the same manner, is available as an optional feature. Weight response of the balance on the 0-100 mg range is 20 μg, with an overall accuracy of ±50 μg.

The sample support rod, one configuration of which is shown in Fig. 29, plugs into a stirrup which rests on the front

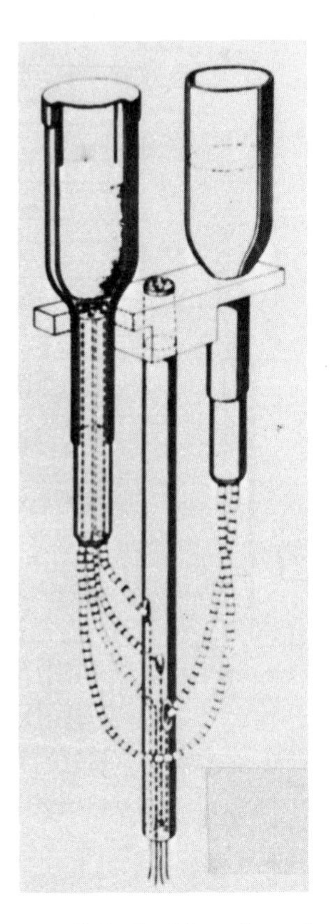

FIG. 29. Dual crucible holder for simultaneous DTA-TGA. Crucibles shown are 8 mm in diameter.

knife edge of the balance beam. Four thermoelectrically compensated copper alloy bands make electrical connections from the thermocouple junctions to the balance housing and the control and measuring circuits.

A unique design feature of the instrument is the gas flow control system, as shown in Fig. 30. For atmospheric or moderate vacuum applications, noncorrosive gases are passed through the balance housing and up into the sample container chamber. The effect of gas flow on the mass measurement is very small,

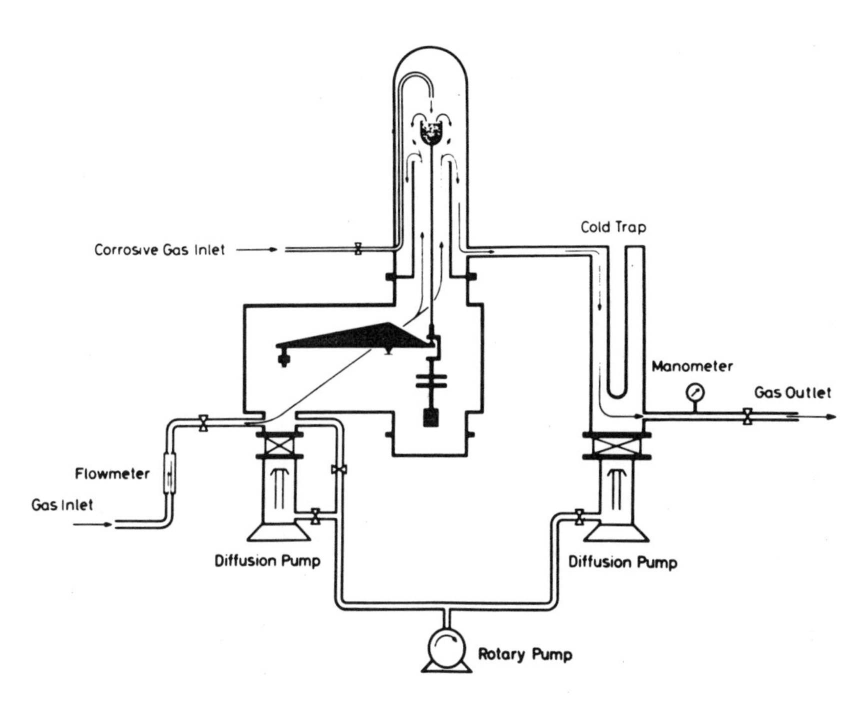

FIG. 30. Gas flow and control system.

being approximately 0.3 mg at the maximum flow-rate of 700 ml/min and 0.1 mg at 160 ml/min. For high vacuum applications, two diffusion pumps, each with a rated capacity of 60 l/sec, are used to evacuate the system. One pump evacuates the balance compartment; the second, the base of the sample tube. A cold trap is used with the second pump, giving the total system a capacity of about 200 l/sec. The low pressure in the system is measured by a built-in ionization gauge located at the base of the sample tube. The ultimate pressure, achieved with a pumpdown time of about 30 min, is on the order of 8×10^{-6} Torr.

The programmer consists of a synchronous motor gear drive, which drives a precision potentiometer to establish a temperature reference voltage. By means of a cam which operates a planetary gear, a correction is added to the gear drive so that the reference voltage corresponds to the standard voltage output curve of a Pt, Pt-10%Rh thermocouple. This voltage is then compared to the actual furnace thermocouple output, and the difference is amplified and used to control the furnace temperature. The accuracy of the temperature program reference voltage is the equivalent of ±0.5°C for all rates from 0.5 to 25°C/min and 25 to 1600°C. The controller is capable of maintaining sample area temperature within ±1.5°C of the required program temperature.

The DTA amplifier has six standard ranges of 1000, 500, 200, 100, 50, and 20 μV per 10-in. recorder span. The curve baseline may be shifted left or right a total of four calibrated steps. A continuous zero adjustment is provided which covers two full spans.

Two standard furnaces are available -- a low-temperature unit with a range up to 1050°C, and a high-temperature unit for operation up to 1600°C.

The basic system consists of the following: a) the recording balance, dual automatic weighing system ranges of 1000 mg and 100 mg, either range recorded simultaneously, or 10-inch full span and 100-inch full span; b) balance stand and cabinet containing the vacuum electrical compensation balance mounted in a double-walled stainless steel vacuum tank, constant temperature water bath for control of the balance temperature, water-cooled heat protection shield and collar and gas flow control system (consisting of a precision needle valve inlet, flowmeter, precision micromanometer, stainless steel cold trap and dual rotary gas ballast vacuum pump), and TGA crucible support stick with Pt, Pt-10%Rh temperature measurement and control thermocouple; c) electronic cabinet containing the electrical weight measurement panel, the temperature program unit, the furnace controller, 11-inch 12-channel multipoint recorder, and power distribution panel; and d) complete installation by manufacturer technicians.

SPECIFICATIONS

Substitution balance, electrically compensated:

Weighing range (incl. crucible weight): 0-16 g.

Tare compensation (electrical): Continuous - 0.1 g; in 0.1 g steps - 0.9 g.

Tare reserve (for weighing below balance): 25.0 g.

Capacity: 42.0 g.

Built-in-switch weights: Weight set - 15.9 g; 1 dialing step - 100 mg; accuracy - 0.1 mg.

Taring weights: 1 x 5 g and 2 x 10 g.

Calibration weights: 0.1 g and 1 g.

Electrical weight detection:

Two ranges: As general diagram - 0-1000 mg and 0-100 mg. As expanded diagram automatically dialled in 10 steps: 1 step (full scale) - =100 mg and =10 mg.

Precision (standard deviation): ±0.5 mg and ±0.05 mg.

Response value: 0.2 mg and 0.02 mg.

Accuracy: General diagram - ±5 mg and 0.5 mg; 1 scale division (2.5 mm 1/10 in.) - =10 mg and =1 mg. Expanded diagram - ±0.5 mg and ±0.05 mg; 1 scale division (2.5 mm 1/10 in.) - =1 mg and =0.1 mg.

Readability: General diagram - 2 mg and 0.2 mg. Expanded diagram - 0.2 mg and 0.02 mg.

Furnace installation:

Temperature program: Range - 25-1600°C. Heating and cooling rates - 0.5, 1, 1.5, 2, 4, 6, and 8. Selectable in steps of rapid increase rates - 10°C/min, 15°, and 25°. Accuracy of temperature gradient - ±10 V = +1°C. Digital temperature indication (calibrated for Pt/Rh 10%-Pt reference temperature 25°C - 1 digit = 1°C. Automatic preselection of constant temperature and cut-off temperature control accuracy - 1 digit = 1°C.

Furnace data: Low-temperature furnace: range 25-1000°C, consumption 1.5 kW. High-temperature furnace: range 25-1600°C, consumption 4.5 kW.

Temperature regulation: Accuracy ±1.5°C.

Temperature signal: Range 0 = 10 mV; automatic dialling 10-20 mV; accuracy ±25 μV.

Extreme low-temperature measurement (e.g., for BET):

Depends on coolant, e.g., liquid nitrogen 77.5°K.

Vacuum system:

Final vacuum (determined in sample compartment without sample): With rotary and two diffusion pumps - 8 x 10^{-6} Torr. Rotary pump alone - 5 x 10^{-2} Torr.

Vacuum gauges: Aneroid fine vacuum gauge - 0 = 760 Torr. Precision thermocouple gauge - 1 to 2 x 10^{-3} Torr. Penning cold-cathode type ionization vacuum gauge - 5 x 10^{-3} to 2 x 10^{-6} Torr.

External gas supply: Maximum permissible flowrate = 30 l/h. Flowmeter range 0-30 l/h.

DTA equipment:

	Range, μV					
	1st	2nd	3rd	4th	5th	6th
Six ranges	1000	500	200	100	50	20
Four calibrated counter EMF's in steps of	1000	500	200	100	50	20
Continuous zeropoint adjustments	2000	1000	400	200	100	40
Response values	5	2	1	0.5	0.5	0.2

Recording of test values:

12-channel recorder: Printing interval - 2.3 sec. Chart speed variable between 2 and 12 in./h = 50.8 and 304.8

mm/h. Width of chart - 11 in. = 279.4 mm. Scale divisions from -5 to 105. 100 divisions = 10 in. = 254 mm.

External connections:

	Low-temperature furnace	High and low temperature furnace
Voltages	3 x 380 V, 50 Hz + zero conductor	3 x 380 V, 50 Hz + zero conductor
	3 x 415 V, 50 Hz + zero conductor	3 x 415 V, 50 Hz + zero conductor
	3 x 208 V, 60 Hz	3 x 208 V, 60 Hz
	3 x 240 V, 60 Hz	3 x 240 V, 60 Hz
Amperage max.	17 A	25 A
Consumption	4 kVA	7 kVA

Water supply:

Pressure: 2-4 atm (30-60 psi).

Consumption: approx. 75 l/h (20 gal/h).

DTG equipment:

Four sensitivities for the 1-g range: 250 mg/min, 100 mg/min, 50 mg/min, and 25 mg/min.

Four sensitivities for the 0.1-g range: 50 mg/min, 20 mg/min, 10 mg/min, and 5 mg/min.

Calibrated counter EMF's in 4 steps (+ twice full chart width). Continuous zero-point adjustment over full chart width.

Time constant: Approximately 4 to 10 sec depending on range. Accuracy: ±5% of full scale. Linearity: ±1%. Response value: ±0.5% of the range.

Balance:

Weighing system: Substitution balance, top- or bottom- loading, with electrical compensating system.

Damping: Electrical damping device.

Knife-edges and bearing planes: Synthetic sapphire.

Built-in weight set and calibrating weights: Nonmagnetic chrome-nickel steel, density 7.77 g/cm^3.

Taring weights: Nonmagnetic chrome-nickel steel.

Crucible support rod: Alumina (with built-in thermocouples).

Sample compartment cover: Alumina or quartz. Outside diameter - 50 mm (2 in.); height - 270 mm (10 1/2 in.).

Case: Chrome-nickel steel.

Hanger for weighing below balance: Built-in.

Furnace and temperature detection:

Furnace arrangement: Vertical, above the balance.

Low-temperature furnace: Kanthal wire coils, bifilar.

High-temperature furnace: Super-Kanthal elements, bifilar.

Thermocouple: Pt-Rh 10%-Pt.

Electronic cabinet:

Rack to incorporate various 19 in. standard panels: Sheet steel.

Optional Equipment:

1) Automatic weighing range, 0-10 mg.
2) High vacuum installation.
3) DTA amplifier installation.
4) Derivative computer installation.
5) Temperature signal amplifier installation.
6) Middle-range furnace installation, 25-1000°C.
7) High-temperature furnace installation, 25-1600°C.
8) High-temperature stainless steel and Al_2O_3 corrosive gas inlet installation.

Accessories:

1) Macro TGA/DTA sample holder stick.
2) Micro welded cup TGA/DTA sample holder stick.
3) Micro TGA-DTA sample holder stick for high vacuum.

4) Middle-range temperature, corrosive-gas inlet quartz furnace, 25-1000°C.

5) Water vapor furnace system.

6) Various sample crucibles of Al_2O_3, Pt, etc.

Thermoanalyzer 2

The Mettler Thermoanalyzer 2, as shown in Fig. 31, is a modified version of the Thermoanalyzer previously described. Although more compact in size, it offers the same advantages of simultaneous TGA/DTG/DTA on a simple sample. All analyses can be carried out in an accurately defined and controlled gaseous atmosphere, even in that of a corrosive nature, or under vacuum. The unit is designed as a tabletop laboratory instrument.

The weighing system is housed in a high vacuum-tight cell. By using an inert carrier gas, it is possible to work with gases of a corrosive nature without damage to the weighing system. The temperature programmer has 26 heating and cooling rates, from 1° to 100°C/min. Or, the system may be operated isothermally at any temperature from 25° to 1000°C. Cycling between two preselected limits is also possible to repeat reversible processes.

The specifications of the unit, with the exception of the high temperature furnace and high vacuum system, are similar to those for the Mettler Thermoanalyzer previously discussed.

DTA 2000 System

This new DTA system fulfills the highest requirements with respect to calorimetric sensitivity and precision as well as accuracy of temperature measurement. It is a modular unit system, hence, flexible and expandable. Accurate temperature control is achieved with the aid of the TA30 program selector, TA32 temperature controller, and TA31 power amplifier. The

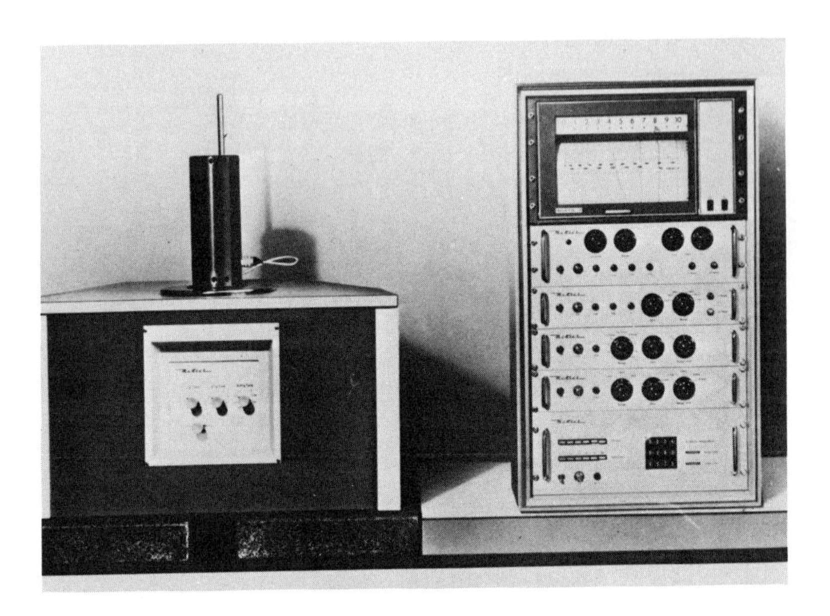

FIG. 31. Mettler Thermoanalyzer 2.

high accuracy of the measuring cell temperature is assured by a platinum resistance sensor. The temperature regulator has upper and lower limiting temperatures, which can be digitally selected for routine operation. Automatic temperature recording markers may be activated or deactivated as desired and are superimposed on the DTA curve every 1°C or 10°C.

The system is illustrated in Fig. 32. The heart of the system is the TA10 DTA cell, which is designed for the temperature range of -20 to 500°C. Differential temperature is detected by a special glass disk sensor which contains five vapor-deposited thermocouples. Exact positioning of the crucibles is assured by centering holes on the measuring head and guide pins on the crucibles. The measuring head has connections for vacuum and purging gas.

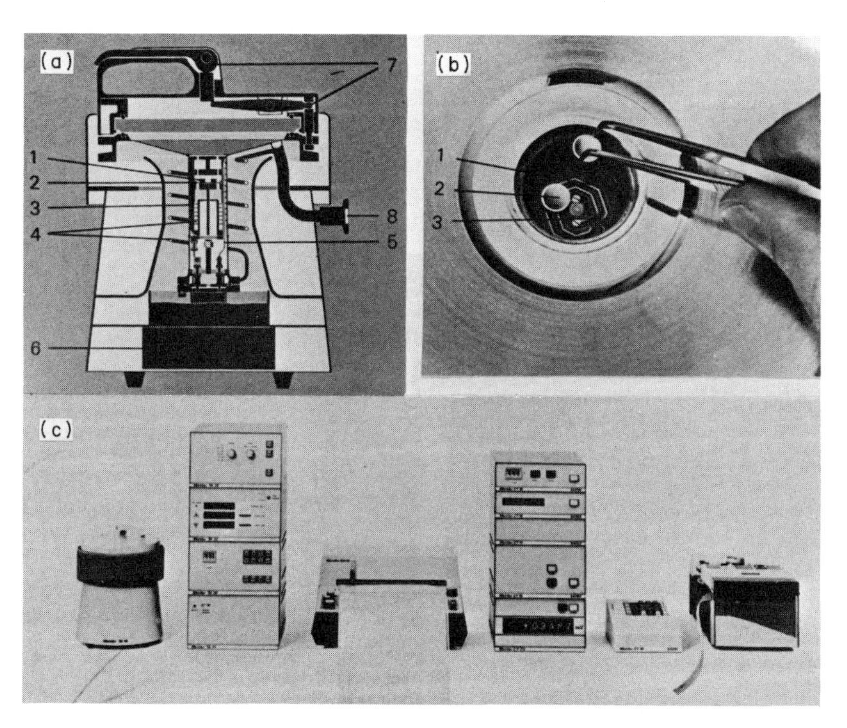

FIG. 32. Mettler DTA 2000 system; (a) DTA cell; (b) measuring sensor; (c) complete system with data transfer system.

SPECIFICATIONS

TA10 DTA Cell

Temperature range: -20° to 500°C.

Calorimetric sensitivity: $\frac{60\ \mu V}{mcal/s}$

Calorimetric reproducibility: ±0.5%.

ΔT measuring sensor (fivefold thermopile): Al/Ni; 115 μV/°C.

Standard crucible (Al, sealable): 40 μℓ.

High-pressure crucible (V2A, up to 100 atm): 370 μℓ.

Purging gas conditioning: Max. 20 ml/min.

Pressure range: 760 to 10^{-3} Torr.

TA20 DTA Amplifier

Measuring ranges: 20, 50, 100, 200, 500, and 1000 μV.

Outputs: L and II (recorder) - 100 mV/measuring range; III (DVM) - 1 V/measuring range.

Zero-point shift (outputs I and II): ±1.2 measuring ranges.

Calibrated EMF suppression, automatically switched (output I): ±3 measuring ranges.

Noise level (0.01 to 1 Hz): $\lesssim$0.1 μVpp.

TA30 Program Selector

Heating and cooling rate: 0.1 to 29.9°C/min.

Digitally adjustable in steps of 0.1°C/min

Rapid heating: 100°C/min.

Temperature program: heating or cooling - automatic isothermal hold; heating or cooling - automatic shut-off; automatic cycling between preset limits.

Temperature markers, on or off (outputs I and II): 1 marker per 1°C or 10°C.

TA31 Power Amplifier

Heater power: max. 500 W.

TA32 Temperature Controller

Accuracy of furnace temperature: $<300°C \pm 0.3°C$ and $> 300°C \pm 0.5°C$.

Reproducibility of furnace temperature: ±0.1°C.

Linearity of heating and cooling: ±0.2%.

Pre-selectable temperature limits (adjustable over entire temperature range): upper/lower.

Data Transfer System

In routine analyses, the connection of a data transfer system offers the possibility of accurate evaluation of the DTA curves in minutes. The Mettler data transfer system, as illustrated in Fig. 33, permits the transmission of the measured

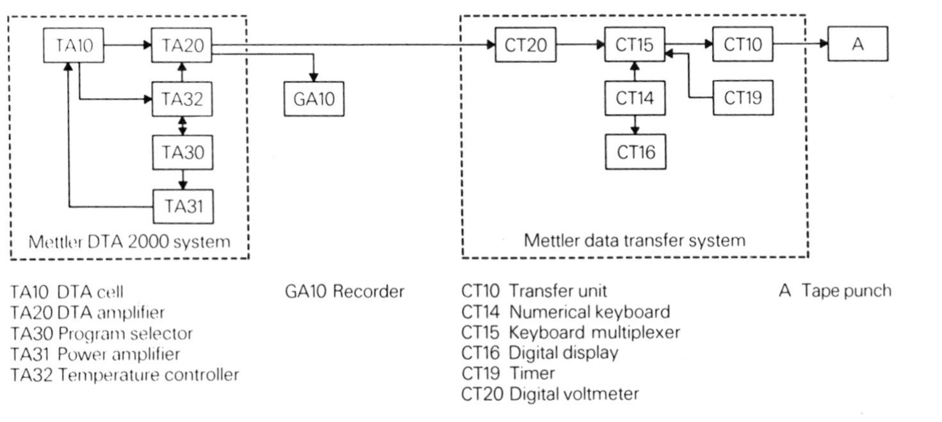

FIG. 33. Mettler data transfer system.

values and apparatus constants to punched tape, for evaluation in off-line operation.

In purity analysis, the contamination in mole-per cent, the enthalpy of fusion, the melting point depression, and the melting point of the pure substance can be determined rapidly with high accuracy.

NETZSCH GERATEBAU GMBH

Differential Thermo Analyzer 404

Two models of DTA instruments are available from this manufacturer, the high temperature Model 404 and the low temperature Model 404T. Both models use the large number of sample holders ("measuring heads") that are available.

The Model 404, as shown in Fig. 34, consists of the following components: a) Temperature Control Unit 406 for furnace temperature control according to a preset program, with ten heating rates; b) furnace frame with furnace and one measuring head; c) Recording Unit for the simultaneous recording of the temperature and temperature difference curves (A choice can be made between the electric compensation recorder and a photographic recording device. The latter permits ten recording speeds, while the former has only three different chart speeds.); and d) Power Unit with intermediate transformer for the heating of the furnace in accordance with the regulating impulses of the control unit. This power unit can be supplied for various power capacities, depending on the kind of furnace used.

SPECIFICATIONS

Temperature Control Unit 406

Control accuracy: ±0.5%.

Power required: 0.5-3.5 kW.

Interchangeable cam disk.

Four programs can be selected.

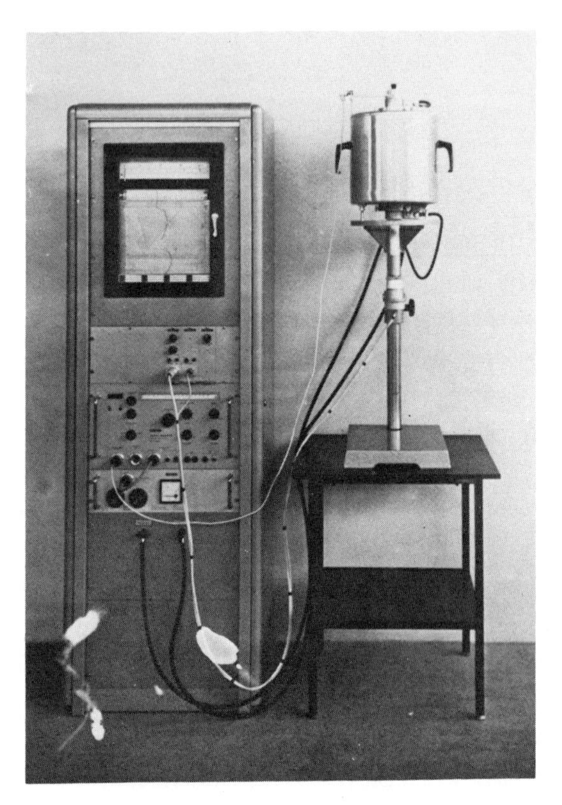

FIG. 34. Netzsch Differential Thermo Analyzer 404.

Recorder

Electrical recorder: maximum sensitivity ±50 μV.

Photographic recorder: maximum sensitivity ±100 μV.

Smallest recordable temperature difference between 0.01 and 0.5°C.

Furnace

Kanthal wire heater element for use to 1320°C. Platinum-rhodium element for temperatures to 1550°C.

Differential Thermo Analyzer 404T

This instrument is similar to the Model 404 with the exception of the furnace arrangement. The operating temperature is from -180° to 420°C and features high sensitivity, a measuring block with four cylindrical chambers, and ten heating rates. The furnace and measuring head are precooled with a cooling medium, usually liquid nitrogen, and then programmed for increasing temperature by the temperature programmer.

SPECIFICATIONS

Temperature range: -180° to 420°C.
Thermocouple: Fe-Constantan.
Volume of sample container: 0.3 ml.
Vacuum: to 10^{-4} Torr.
Maximum heating rate: 0.04° to 40°C/min.
Cooling time: 100 min to -160°C.
Linearity of control: ±0.5%.

Sample Holders (Measuring Heads). There are a number of different measuring heads available, each constructed of a specific material for use to a certain maximum temperature. Materials of construction range from alumina to platinum. Schematic diagrams of the measuring heads are shown in Fig. 35. All heads are interchangeable and can be used with each system. Data on the available measuring heads are presented in Table 5. Selection of the most appropriate measuring head may be made by use of the data in Table 6.

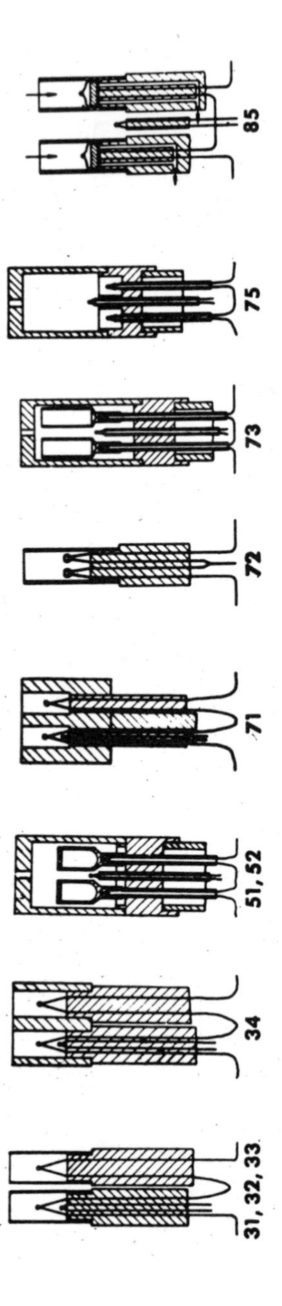

FIG. 35. Netzsch DTA measuring heads.

TABLE 5

Data on Netzsch Measuring Heads

Measuring head	Type	Thermocouples	Capacity, ml	Max. Temp., °C
Standard	31	Pt/Rh-Pt	1	1600
Vacuum	32	Pt/Rh-Pt	1	1600
Protective gas type	33	Pt/Rh-Pt	1	1600
Standard Ni block	34	Pt/Rh-Pt	1	1000
DDK	51	Pt/Rh-Pt	0.25	1000
Block	71	Pd-Pt	0.2	1000
Special microhead	72	Pt/Rh-Pt	1	1600
Head for plastics	73	Ni/Cr-Ni	1	450
Head for dust samples	75	Pt/Rh-Pt	0.2	1000
Catalyst head	85	Ni/Cr-Ni	1	900

TABLE 6

Selection of Most Suitable Measuring Head

Application	Sample state	Head type		
		1320°-1550°C	900°C	400°C
Raw material control	powder	31	34	34
Fast analysis	fine powder	--	75	75
Chemical research	melting substances	51	51	51
Mechanism of reactions	small amounts of powder	--	71	71
Catalyst gas reactions	grains	--	85	85
Quantitative analysis	as desired	51	51	51
Calorimetry	as desired	51	51	51
Metals	solid	31,51	31,51	31
Synthetics	fibers, general	--	--	73
Fats, soap, oil	as desired	--	--	73

EDWARD ORTON, JR., CERAMIC FOUNDATION

Thermal Analyzer

The Orton Thermal Analyzer is not a differential thermal analysis instrument, but since it does obtain calorimetric data of a sample it will be included here. This thermal technique, which measures the temperature gradients within the sample, is called direct calorimetric analysis (DCA) and is illustrated in Fig. 36.

In this technique, the cylindrical sample contains a thermocouple in its geometrical center, which provides the signal to record the heating rate. This thermocouple is also differentially connected to a surface or reference thermocouple. An adjustable voltage is applied to the thermocouple signal, which is used to maintain the temperature differential in the sample. A heater in the block provides whatever power is thus needed to maintain this preset differential.

When reactions or changes of state occur within the sample, the uniform temperatures profile is disturbed. The correlation between heating rate and temperature differential, which depended upon diffusivity before the reaction started, is now replaced by a new control mechanism. Well crystallized and pure materi-also exhibit practically isothermal decomposition in this equipment. A run may be completed within 3 hours or may be extended to more than 24 hours if slow heating rates are desired.

The ΔH of reaction may be obtained from the DCA curve if the specific heat Cp is known (either from direct measurement or from other sources) by use of the expression,

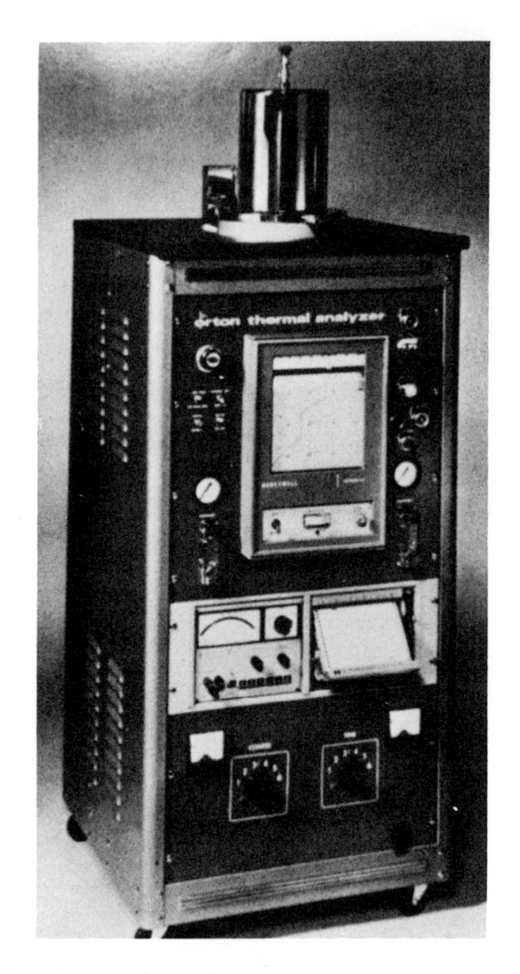

FIG. 36. Orton thermal analyzer.

$$\Delta H = Cp\ (T_2 - T_1),$$

where $(T_2 - T_1)$ is the temperature difference.

The furnace on the instrument has maximum temperature limits of 1200°C (Kanthal A), 1600°C (Pt, 40% Rh), or 1700°C (molybdenum). Samples may be solid or liquid and tests may be conducted in air, vacuum, or one- or two-gas atmospheres. Heating or cooling may be employed.

The instrument finds the greatest application to problems in ceramics technology and to other areas where a knowledge of the sample thermal conductivity or diffusivity is required.

PERKIN-ELMER CORPORATION

The difference between DTA and differential scanning calorimetry (DSC) has previously been discussed. In DSC, as the temperature is programmed, the sample and reference temperatures are continuously maintained at the same level. This is accomplished by closed-loop, negative feedback control of the power supplied to the sample and reference holders. When the sample absorbs or evolves heat, more, or less, power is required by the sample holder to maintain it at the same temperature as the reference holder. This differential power is recorded as the ordinate in DSC. Since power is energy-per-unit-time, the curve represents the rate of energy absorption as a function of sample temperature. On a linear time-base recorder, the peak area represents the heat of the associated thermal transition. The calibration for conversion of area-units to calories is independent of temperature, scanning rate, and the nature of the sample.

Perkin-Elmer Model DSC-2

The Model DSC-2, as illustrated in Fig. 37, is a new model of the differential scanning calorimeter, originally introduced by Perkin-Elmer in 1963. This latest version of the calorimeter, which was first exhibited in March of 1972, contains many improvements over the Model DSC-1B. It extends the temperature range by 225°C to a maximum temperature of 725°C, and by 75°C at the lower end, to a low temperature of -175°C. The noise level has been reduced by a factor of ten over the range

FIG. 37. Perkin-Elmer Model DSC-2 calorimeter.

permitting calorimetric sensitivities down to 0.1 mCal/sec full scale. Also, the temperature program linearity has been improved by a factor of ten so that the temperature calibration between 50° and 725°C may be adjusted to a linearity of ±1°C. A high resolution digital programmer (30 digital steps per degree) combined with a compensating-filter network provides essentially instantaneous response of sample holder temperature to program changes. Digital temperature readout is within 0.1°C with precise and repeatable temperature limits settable at intervals of 0.1°C. There are eleven heating rates from 0.325°C/min to 320°C/min and the same but independently selectable cooling rates. A variety of analog and digital signal outputs are provided for maximum readout system flexibility and compatibility with acquisition systems for computer analysis of the data.

SPECIFICATIONS

Temperature (abscissa)

Range: 50°C to 725°C (320°K to 1000°K) -175°C to 725°C (100°K to 1000°K) with subambient accessory.

Readout: 7 segment digital display: in °K from 100.0 to 999.9; in °C from 000.0 to 725.0; in °F from 000.0 to 999.9.

Readout accuracy: ±1.0°C.

Precision: ±0.1°C.

Measuring accuracy: ±0.1°C (using calibration correction curve).

Sensing elements: Platinum resistance thermometers.

Temperature Programmer

Heating rates: 0.3125°, 0.625°, 1.25°, 2.5°, 5°, 10°, 20°, 40°, 80°, and 160°C.

Cooling rates: The same but independently selectable.

Controls: "Heat," "Hold," and "Cool" pushbuttons; "Hold," "AutoCool" and "Cycle" mode selector. Digital upper and lower limit selectors settable to 0.1°C.

Differential Power (ordinate)

Range: 0.1, 0.2, 0.5, 1.0, 2.0, 5.0, 10.0, and 20.0 mCal/sec. full scale on 10 mV potentiometric recorder.

Noise (at 700°K): Isothermal, ±0.002 mCal/sec. maximum; scanning, ±0.004 mCal/sec. maximum.

Atmosphere: Nitrogen or Argon static or dynamic purge up to 200 cm^3/min from 0.5 to 3 atm pressure. Helium required in subambient operation with liquid nitrogen coolant.

Dimensions: W = 40 in. (101.6 cm); D = 25 in. (63.5 cm); and H = 19 3/16 in. (48.7 cm).

Weight: 180 lb (81.6 kg) net

Power requirements: 230 VA; 230 V at 1 A, 115 V at 2 A.

RIGAKU-DENKI

Differential Thermal Analyzer

Two models are available, both of which feature desktop mounting. Because of the modular design, attachment units can be selected and connected to the basic unit in order to make the desired measurement for a variety of applications. The Differential Thermal Analyzer, which measure only the DTA curve of the sample, or the Differential Thermal and Thermogravimetric Analyzer, in which both DTA and TG curves are recorded, may be obtained. The former is illustrated in Fig. 38.

The compact DTA instrument uses a small thermal capacity furnace which permits rapid cooling times. Power consumption is one-fifth that of earlier models and the temperature distribution is more uniform.

The modular design accepts different attachments suited to a particular type of sample and application. By preparing two or more sample holders, samples can be run sequentially without waiting for the furnace to cool. Also, by preparing multiple sample holders, measurements in different temperature ranges can be made. Measurements can be made in air, vacuum, or flowing gas atmosphere. A gas chromatograph can be connected for effluent gas analysis.

SPECIFICATIONS

Recorder

Recording, 2-pen type: Pen No. 1 - DTA; Pen No. 2 - temperature.

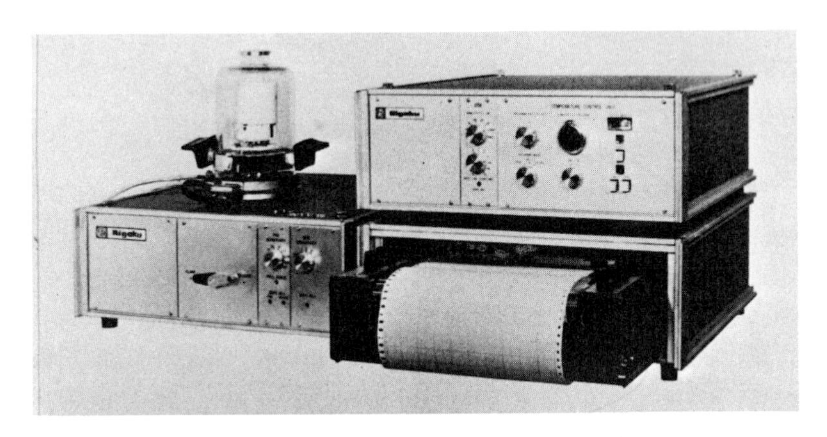

FIG. 38. Rigaku differential thermal and thermogravimetric analyzer.

Pen speed: Within 1 sec/full scale.

Width of chart paper: 250 mm.

Chart speed: 2, 4, 8, and 16 mm/min.

Specimen holder (with electric furnace)

Standard type:

Temperature range: Room to 1100°C.

Quantity of specimen: 0.5 cm^3 (10 to 500 mg).

Thermocouple: Chromel-Alumel.

Specimen ampoule: Transparent quartz.

Heater: Ferrochrome wire.

High temperature type:

Temperature range: Room to 1500°C.

Quantity of specimen: 0.5 cm^3.

Thermocouple: Pt, Pt-13%Rh.

Specimen ampoule: Pt, Rh alloy.

Heater: Pt, Rh wire.

Micro specimen type:

Temperature range: Room to 1100°C.

Quantity of specimen: 0.05 cm^3 (0° to 50 mg).

Thermocouple: Chromel-Alumel.

Specimen ampoule: Aluminum.

Heater: Ferrochrome wire.

Continuous high-low temperature type:

Temperature range: -150° to 800°C.

Quantity of specimen: 0.5 cm^3.

Thermocouple: Chromel-Alumel.

Specimen ampoule: Transparent quartz.

Heater: Ferrochrome wire.

Temperature controller

Continuous high-low temperature type:

Control system: Programmed P.I.D. control.

Heating and cooling rate: 1°, 3°, 5°, 10°, 15° and 20°C/min.

Temperature range: -150° to 1500°C.

Measurement modes: Heat, hold, and cool.

Accuracy: to 1100°C within ±0.2°C; to 1500°C within ±0.5°C.

Program control type:

Control system: Programmed P.I.D. control.

Heating and cooling rate: 1°, 3°, 5°, 10°, 15° and 20°C/min.

Temperature range: Room to 1500°C.

Measurement modes: Heat, hold, and cool.

Accuracy: to 1100°C within ±0.2°C; to 1500°C within ±0.5°C.

Differential amplifier unit

Type: Transistorized dc voltage amplifier

Amplifier range: ±10, ±25, ±50, ±100, ±250, ±600, and ±1000 μV/full scale.

Linearity: ±0.5%.

Accuracy: ±1%.

Atmosphere control unit

Vacuum gauge: Pirani gauge.

Gas flow gauge: 0 to 1000 ml/min.

Valves: One each for vacuum, gas flow, and leak.

Cooling unit

Coolant: Liquid nitrogen.

Coolant capacity: 10 liters.

Heater for generating coolant gas: Ferrochrome wire.

Desktop TG-DTA Unit

This instrument is a compact desktop thermogravimetric and differential thermal analysis unit designed for simultaneous measurements of both of these parameters. It features the capability of using small sample sizes, indicates the mass-loss in percentage mass-loss as well as in milligrams, and features rapid sample changing. It is modular in design which permits different attachments to be added at any time, depending upon the research requirements.

SPECIFICATIONS

Balance and thermogravimetric circuit

System: Suspension-type electric balance.

Quantity of specimen: 0 to 500 mg.

Accuracy: 10 μg.

Weighing range: 1, 2, 5, 10, 20, 50, 100, 200, and 500 mg/full scale.

Recorder: Three-pen type with a pen speed of 1 sec/full scale. Chart speeds of 2.5, 5, 10, 20, 40, and 80 mm/min.

DTA amplifier: Amplifier range of ±10, 25, 50, 100, 250, 500, and 1000 μV/full scale.

Temperature programmer: SCR-PID type with modes of heat, hold, cool, and recycle. Heating rates of 0.625°, 1.25°, 2.5°, 5°, 10°, 20°, 40°, 80°, and 160°C/min.

Furnace: Range from 25° to 1100° or 1500°C; heater element is either ferrochrome or Pt/Rh wire.

Optional attachments include: a) gas-flow attachment which prevents corrosive gases from entering the balance chamber; b) derivative TG circuit for obtaining the derivative of the mass-change curve.

SETARAM
(Societé d'Etude d'Automatisation de Regulation
et d'Appareils de Mesures)

Micro Differential Thermal Analyzer Model M-4

Micro DTA instruments, as pioneered by C. Mazieres, provide for the study of samples ranging in mass from several micrograms to 50 mg. The Model M-4 employs many of the design concepts of Mazieres and yet provides high resolution and temperature accuracy.

The instrument is modular in construction and consists of the probe, furnace mount, measuring circuits, atmosphere control, and probe cooling circuits mounted on a chromium-plated table on top of the cabinet. The temperature programmer, power supply, and recorder are mounted externally. The instrument is shown in Fig. 39.

Three interchangeable probes are provided: a) micro probe of 0.6 μl capacity, for highest resolution; b) semi-micro probe of 6 μl capacity; and c) semi-micro probe of 25 μl capacity.

Each probe has a measuring head equipped with three hollow thermocouples (two for ΔT and one for T measurements) mounted inside a platinum metal chamber. Each thermocouple consists of a platinum microcrucible to which the two wires are welded. Sample cups fit inside the semi-micro probe crucibles. Standard thermocouples are of Platinel II and cover the temperature range of -140° to 200°C. Probes for higher temperature studies are equipped with Pt, Pt-Rh thermocouples.

FIG. 39. SETARAM micro differential thermal analyzer, Model M-4.

According to the manufacturer's literature, strictly linear temperature programming is not essential in micro DTA, since the efficiency of the probe is very little affected by small variations of heating rate. Also, quantitative measurements are made by integration of peaks rather than by measurement of peak height. The open loop programming of the Model M-4, however, gives nearly linear heating rates over most of the temperature range. Four heating rates are available at 2°, 4°, 10°, and 20°C/min. The reproducibility of this programmer is claimed to be excellent if the line voltage remains constant.

The standard furnace, Model F-10, has a Kanthal heater element wound on an alumina core. Its maximum power consumption is 600 W. The furnace mount can be moved aside to clear the probe and in this position it can be cooled by a built-in fan. Cooling time from 1100°C to ambient temperature is approximately 30 min. The high temperature furnace, Model F-20, is equipped with a Pt-20%Rh heater element.

The probe chamber can be flushed with a dynamic gas atmosphere. Gas flow can be reversed for collecting effluents without excessive dilution. The system will hold a vacuum of greater than 5×10^{-3} Torr.

A dual-pen recorder, along with a dc amplifier in the channel used for ΔT recording, is recommended (but not furnished). Simple pen recorders with proper channel switching for ΔT and T signals can also be used. Full-scale sensitivities for the ΔT channel should range from 20 to 250 μV.

SPECIFICATIONS

Temperature range: 25° to 1200°C - standard furnace; 25° to 1600°C - Pt-Rh furnace; -160° to 350°C - low temperature accessory.

Thermocouples: High sensitivity: Platinel II; high temperature: Pt-Rh (6-30%); low temperature: gold-cobalt/copper.

Maximum sample capacity: Microprobe - 0.6 μl; semimicro probe - 6 μl; semimicro probe - 25 μl.

Heating rates: 2°, 4°, 10°, and 20°C/min.

Atmosphere control: Dynamic gas vacuum to 5×10^{-3} Torr.

Utilities: Power requirements - 600 W, 115 or 230 VAC. Cooling water - 2 ft^3/hr.

Dimensions (without recorder): 19 x 17 in. table area.

Weight: 49 lb.

Calvet Microcalorimeter

The Calvet Microcalorimeter is a conduction-type apparatus based on the measurement of heat flow. The temperature detection system is a thermal "fluxmeter" consisting of a set of thermocouples in series, welded alternatively to the wall of the cell containing the sample and to the isothermal metal block. A thin dielectric electrically insulates the junctions from the wall but maintains good thermal conductivity. The fluxmeter has two functions:

1. To transfer the heat of the sample to the block (or vice versa). This transfer is continuously proportional to the temperature difference between the sample cell and the block.

2. To supply an electromotive force proportional to this difference in temperature and the thermal flux. The values of the flux and of the EMF are always identical even if the thermal transition varies over a long period of time.

Two models of the instrument are available, as shown in Fig. 40. The Standard model is for use in the temperature range from 20° to 200°C and has a sensitivity of about 10 μW. The High Temperature model is for use up to 800°C with a sensitivity of about 50 μW. Four different types of sample cells are available for the Standard model which vary from a volume of 15 to 100 ml in different configurations.

C.R.M.T. Calorimeter

The C.R.M.T. Calorimeter is a heat-flow instrument permitting measurements to be made at a constant temperature in the temperature range of 20° to 80°C. It is designed to study the heats of reaction between liquid-liquid or liquid-solid mixtures and also heats of combustion and specific heat determinations. Its simplicity and rapidity of operation make it suitable for basic research and for routine studies in calorimetry; it is said to be useful for traing purposes in this type of calorimetry.

The principle of operation of the instrument is as follows: The heat evolved or absorbed in the sample cell results in a small temperature change (from 0.1° to 0.001°C) which triggers the heat flow between the cell and the block. This heat flow is detected and recorded and represents the amount of heat exchanged; it provides a thermal history of the reaction as a function of time, and integrates the peak area, giving the total quantity of heat evolved or absorbed.

The calorimeter is shown in Fig. 41. Two cell models are available, a standard cell model (17 x 80 mm) and a large cell model (35 x 100 mm). Temperature changes and heat flow are measured with a multijunction thermopile which is arranged radially about the cell. A novel feature of the calorimeter is the oscillating support equipped with a multispeed mechanical drive. This facilitates the measurement of heats of reaction from various types of mixtures.

Accessories for the calorimeter include a Joule effect calibration cell, a mixture cell, and a microbomb for heat of combustion studies.

Simultaneous DTA-TG Instruments

The multipurpose DTA-TG instruments are described in Part II under Thermobalances, page 207.

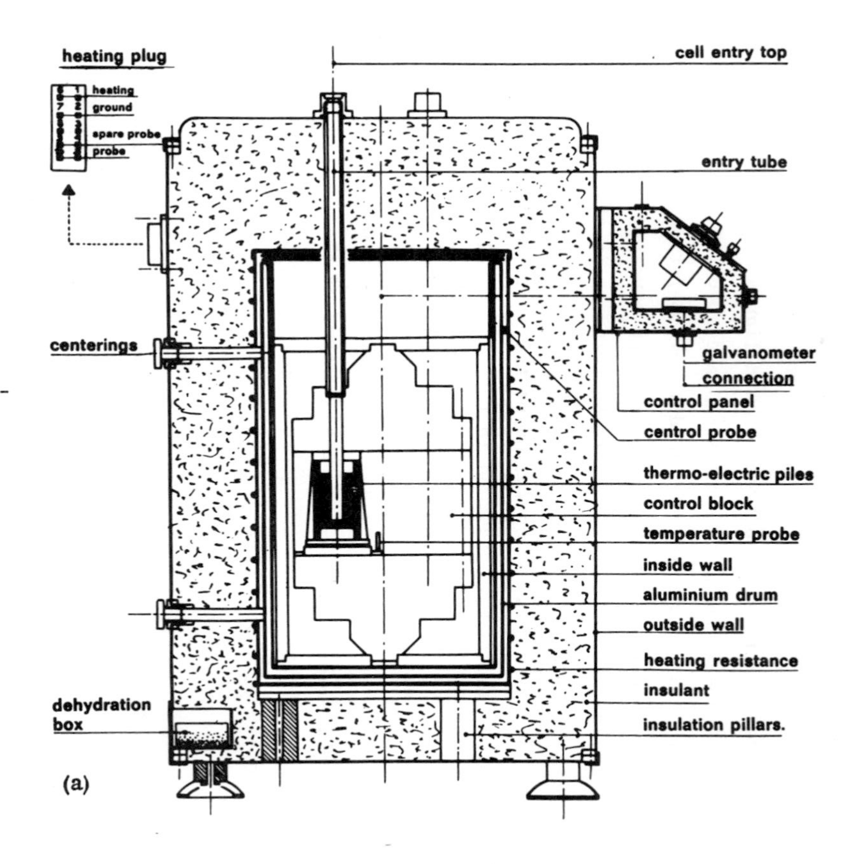

FIG. 40. The Calvet Microcalorimeter: (a) standard model; (b) high temperature model; (c) photograph of apparatus.

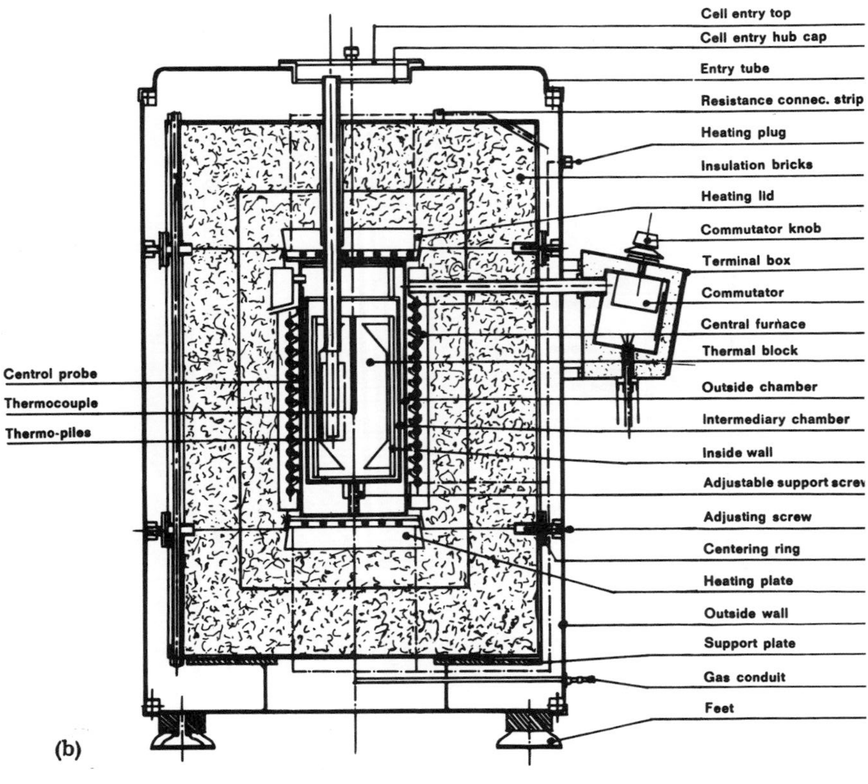

Cell entry top
Cell entry hub cap
Entry tube
Resistance connec. strip
Heating plug
Insulation bricks
Heating lid
Commutator knob
Terminal box
Commutator
Central furnace
Thermal block
Outside chamber
Intermediary chamber
Inside wall
Adjustable support screw
Adjusting screw
Centering ring
Heating plate
Outside wall
Support plate
Gas conduit
Feet
Centrol probe
Thermocouple
Thermo-piles

(b)

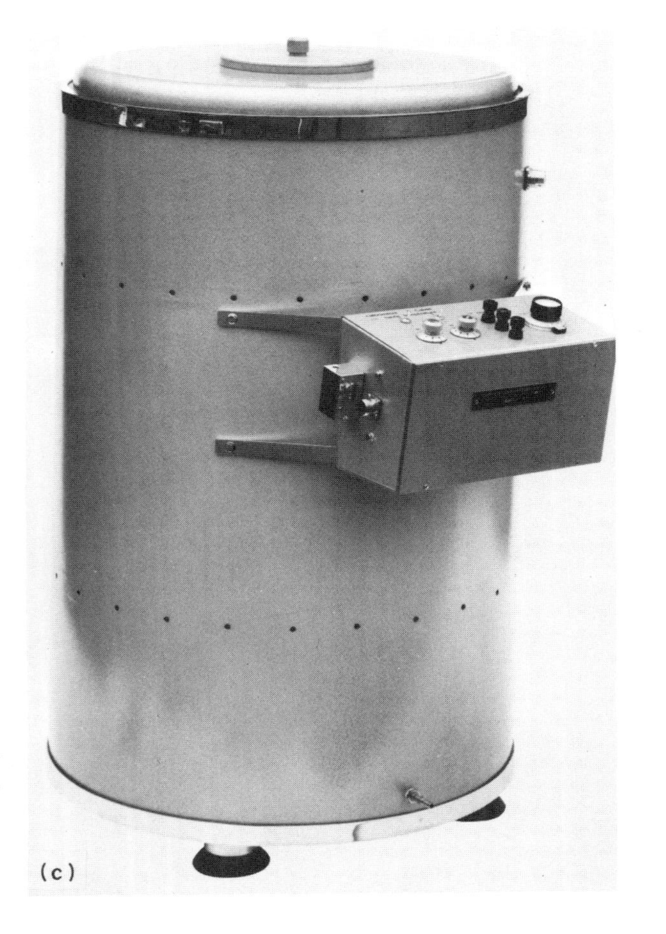

(c)

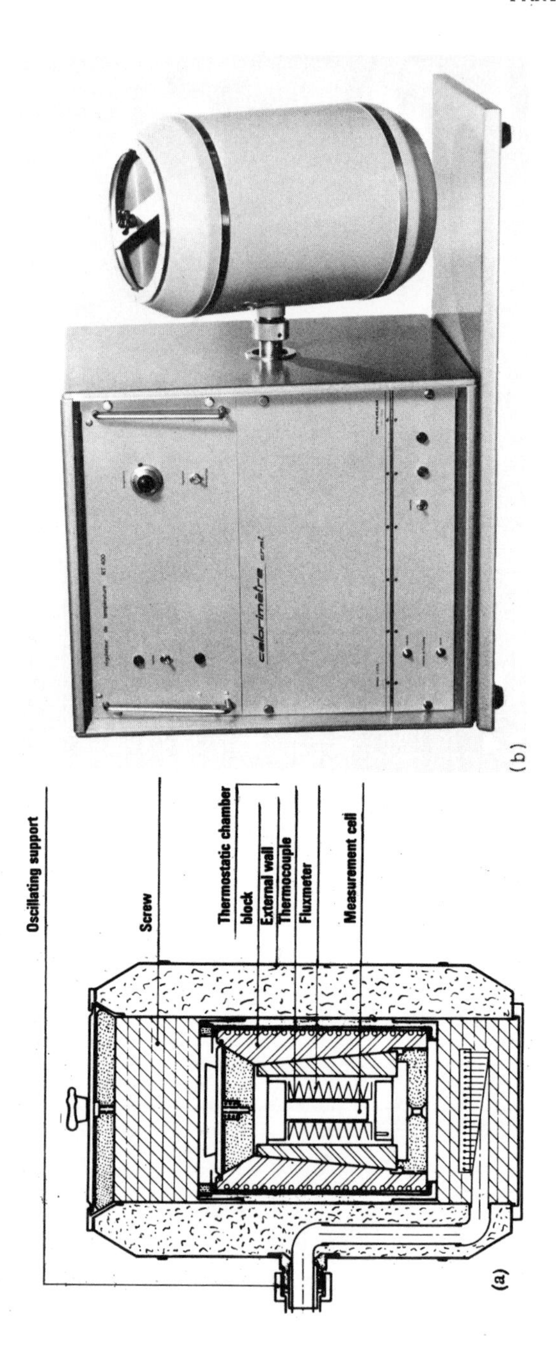

FIG. 41. C.R.M.T. calorimeter: (a) schematic diagram; (b) photograph.

SHIMADZU SEISAKUSHO

Micro DTA System DTA-20B

The Model 20B is shown in Fig. 42. This microtype DTA system features a rather unique type of sample holder. The samples are placed in small cups which rest on a dumbbell-shaped thermoelectric disk. This type of detector has high sensitivity and a good baseline under most conditions.

The sample holder, which permits studies between 25°C and 600°C, consists of the dumbbell detector, furnace, baseline stabilizer, and holder base. Furnace design is such that the temperature distribution is very uniform throughout the sample area. A horizontal, straight baseline is achieved by use of the baseline stabilizer. Any type of sample can be studied, such as solids, liquids, powders, thin films, crystals, chips, pastes, or fibers. Samples are contained in 0.04-ml aluminum pans, or, if they react with this material, 0.06-ml platinum pans. A hermetically sealed sample pan is available for highly volatile samples. Measurements can be made in air, inert or reactive gases at atmospheric pressure, or in a vacuum. Flowrates can be varied from 0 to 60 ml/min; low pressure operation is possible down to 10 Torr.

SPECIFICATIONS

Temperature:

Measurement range: 25° to 600°C. Three-step selection of 15, 25, and 50 mV. Upper and lower temperature limiter incorporated.

FIG. 42. Shimadzu micro DTA system DTA-20B.

Heating rate: 11 rates of 0.5°, 1°, 2°, 5°, 7.5°, 10°, 15°, 20°, 30°, 40° and 80°C/min.

Programming mode: Up, hold, down, cycle.

Control system: PID-SCR; automatic, manual-automatic, and manual.

Temperature difference:

Measurement range: 7 step selection of ±10, ±25, ±50, ±100, ±250, ±500, and ±1000 μV/full scale.

Sample:

Weight: 0.1 - 40 mg.

Volume: Less than 0.04 ml (Al cell); less than 0.06 ml (Pt cell).

State: Solid and liquid. Powder, sheet, clip, paste, fiber (with optional accessory), etc.

Atmosphere: Air, active gas, inactive gas (0 - 60 ml/min); vacuum down to 10^{-2} Torr.

Detection:

Detector: Dumbbell-type detector.

Sensitivity: 0.2 μV (C-A thermocouple).

Recording:

Recorder: 2-pen transistorized potentiometer.

Chart width: 250 mm.

Chart speed: 6 speeds of 1.25, 2.5, 5, 10, 20, and 40 mm/min.

Dimensions:

DT-20B: 487 x 375 x 233 mm.

R-202: 450 x 390 x 152 mm.

MDM-20: 150 x 275 mm.

Weights:

DT-20B: 18 kg.

R-202: 15 kg.

MDM-20: 4 kg.

Optimal accessories:

Cooling blower: Employed when cooling the furnace after measurement. Cooling capacity: 15 min from 600°C to 50°C.

Sample Crimper and Sealer: The contact of the sample and the cell is incomplete with fibrous samples and reproducible data cannot be obtained. However, when the sample is crimped with the Sample Crimper, good reproducible data can be obtained at high sensitivity. In the case of a sublimating or evaporating sample, accurate measurement can be performed by sealing the sample in the cell to eliminate the interference of sublimation and evaporation. The Sample Crimper can be modified into a Sample Sealer simply by changing the dies.

Type DT-2B Differential Thermal Analysis Apparatus

The Type DT-2B DTA apparatus is illustrated in Fig. 43, with a schematic diagram given in Fig. 44. This unit is a highly sophisticated DTA apparatus. It is essentially an updated version of the successful Type DT-2A instrument, the main improvements being in the recording system and the temperature programmer. Both of these now employ completely solid-state electronics. Features of the Type DT-2B are the following:

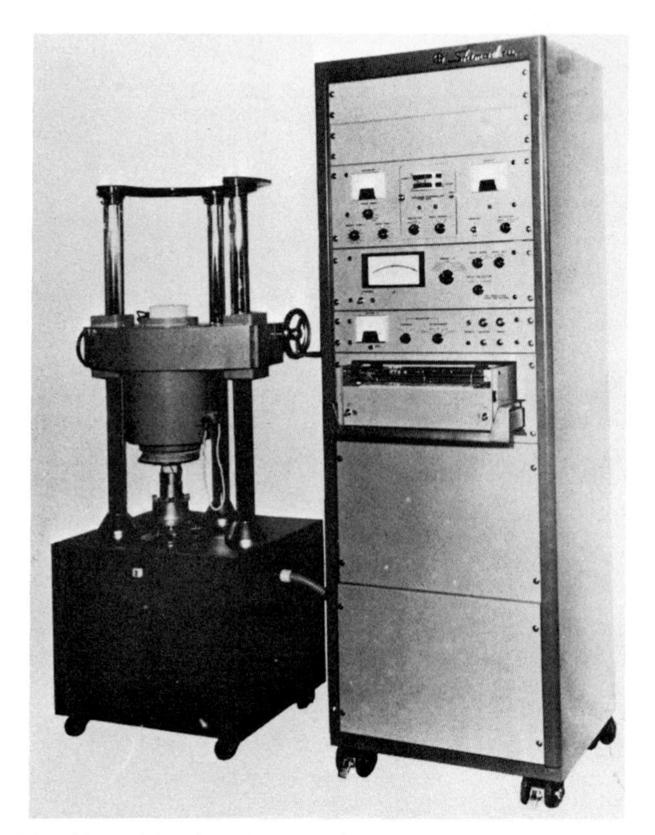

FIG. 43. Shimadzu Type DT-2B DTA apparatus.

1) all solid-state three-pen strip-chart recorder, 2) fully electronic compact temperature programmer, 3) high sensitivity, all solid-state dc microvolt amplifier, 4) furnace of unique design to prevent induction noise at high temperature and for evenness of temperature distribution, and 5) high sensitivity, stability, and reproducible sample holder configuration.

The furnace section (Fig. 44) is a vertical tube furnace cut in two with internal helices and a built-in platinum shield net for equalization of heat distribution and elimination of induction noise. In the center of the furnace there is a block

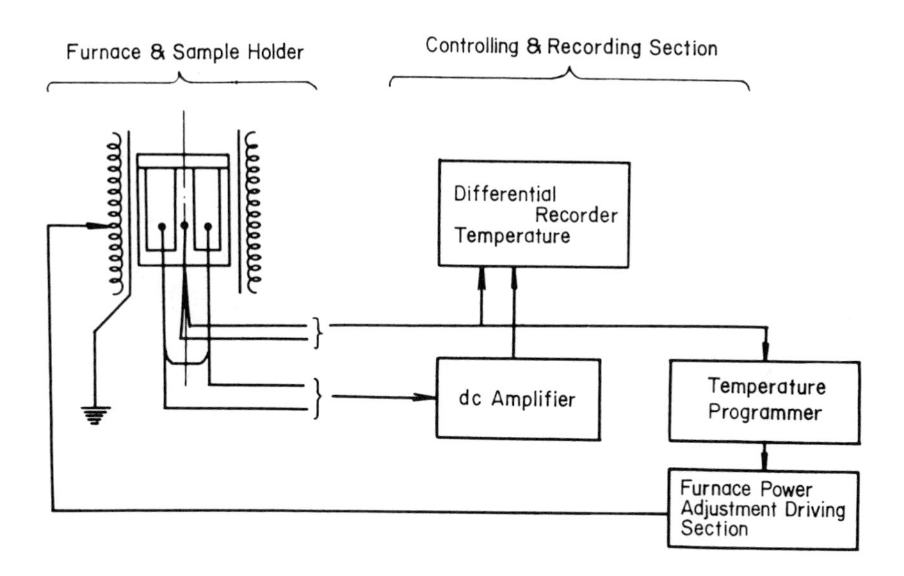

FIG. 44. Schematic diagram of Shimadzu Type DT-2B DTA apparatus.

which is temperature-controlled by the programmer. In programming, temperature of the block is raised or lowered at a desired rate by means of the programmer and various special programming can be made in accordance with requirements. The recorder is a new, all solid state, three-pen unit.

The instrument consists of three units, a) furnace section, b) sample holder, and c) control and recording section. The units in the control and recording section can be used for other thermal accessories, such as TG, dilatometry, and so on.

The standard sample holder consists of a thin platinum alloy cell which is placed in a beryllium oxide block (Fig. 45). For measurements up to 1500°C, the Type BHL-21VB sample holder is employed along with high temperature furnace HF-2.

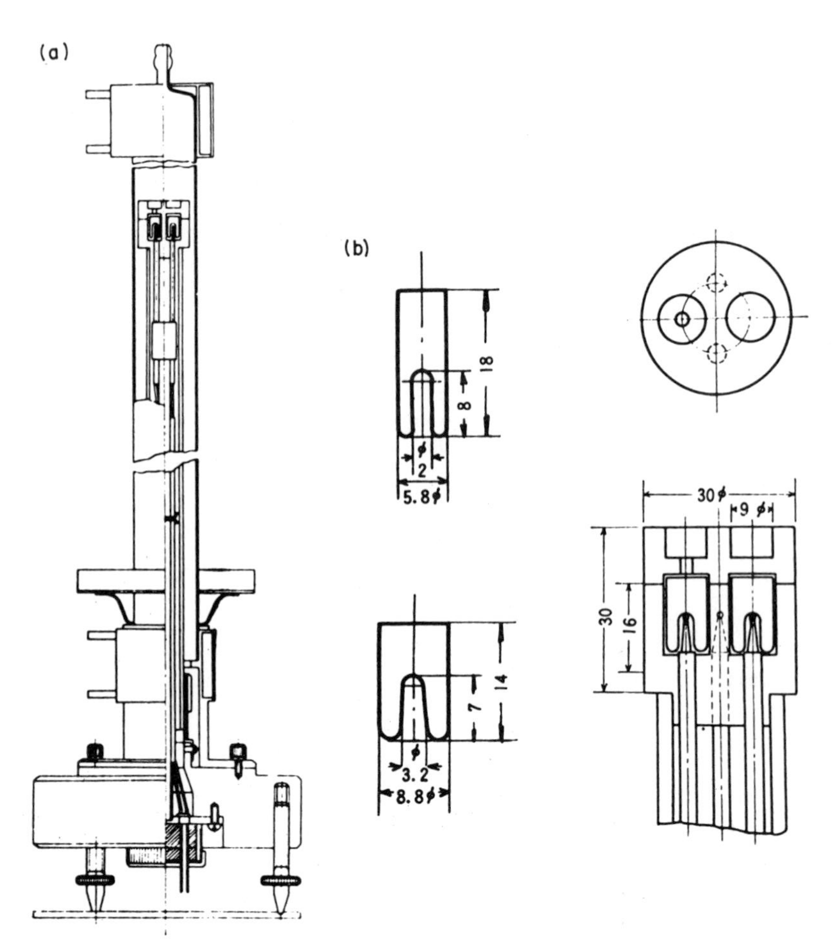

FIG. 45. Shimadzu Type DT-2B DTA apparatus: (a) sample holder Type BHL-21VB; (b) sample cell and block.

A low temperature, high sensitivity unit, Type BHL-22VB, uses Chromel-Alumel rather than Pt, Pt-10%Rh thermocouples.

SPECIFICATIONS

Controller

Temperature range: 25-1500°C; 25-1000°C.

Heating rates: 0.5°, 1°, 2°, 5°, 10°, and 20°C/min.

Power requirements: 100 V ac, 50-60 Hz.

Amplifier

Input range: ±10 to ±5000 μV in 9 steps.

Output: ±10 mV.

Zero point: Continuously variable.

Recorder

Type: 3-pen potentiometric.

Chart width: 250 mm.

Chart speed: 5 or 10 mm/min.

Range: 0 to 1500°C; -150° to 850°C.

Accuracy: ±0.5%.

Furnace

Type: 1) MF-2 Kanthal wire, 1200°C, 1.5 kW. 2) HF-2 Pt, Pt-20%Rh, 1500°C, 2.5 kW.

Dimensions: 560 x 500 x 1400 mm.

Sample Holder

Type: Symmetrical sample well, vertical type.

Block: Beryllium oxide, 30 mm diameter x 30 mm high. Sample well is 9 mm diameter.

Thermocouple: Pt, Pt-10%Rh.

Sample cell: 6 or 9 mm diameter, Pt-Rh alloy.

Overall dimensions of controller and recording section: 575 x 515 x 1715 mm.

STANTON INSTRUMENTS

Standata 6-25 and 5-50

The Standata models are the standard DTA instruments manufactured by Stanton Instruments Ltd. Recently, the newer DTA unit, the Model 67, was introduced. This model is designed for table-top operation.

The Model 6-25 is illustrated in Fig. 46. The instrument is assembled in a floor-mounted cabinet which contains the new variable-rate linear temperature programmer, a new sample holder permitting higher sensitivities, and a choice of three different furnaces for use from -150° to 1500°C. Specifications for the Models 6-25 and 5-50 are given in Table 7.

Stanton Redcroft 67 System

This system is a convenient bench-top unit which employs the modular approach. The Basic 67 module contains the solid-state dc amplifier, which has seven fixed-gain settings from 20 to 1000 μV full-scale deflection. Also contained in the module is a solid-state temperature programmer, which has ten switch-selected heating rates from 1° to 20°C/min, as well as the operating modes of heat, cool, hold, and isothermal. The Basic 67 module with two of the DTA modular furnaces is shown in Fig. 47.

There are four DTA modules available, covering the temperature range from -150° to 1500°C.

TABLE 7

Specifications for Standata Models 6-25 and 5-50

Parameter	Model 6-25	Model 5-50
Temperature range	25° - 1000°C 25° - 1500°C -150° - 450°C	25° - 1000°C 25° - 1500°C -150° - 450°C
Heating rate	1.0° to 20°C/min, continuously variable	
Sample size	1 to 200 mg	1 to 200 mg
T sensitivity	1, 2, 4, 10, 20, and 40 μV/cm	2, 4, 8, 20, and 40 μV/cm
Data presentation	Dual pen, one each for ΔT and T	Single pen, ΔT and T switching
Quantitative measurements	Special recorder equipped with integrator	
Chart speeds	150, 300, 600, 750, 1200, and 1500 mm/hr	150, 300, 600, and 1200 mm/hr
Dimensions	148 x 99 x 59 cm	148 x 99 x 59 cm
Weight	225 kg (475 lb)	225 kg (475 lb)

FIG. 46. Standata Model 6-25 DTA instrument.

SPECIFICATIONS

DTA Module 1B

Temperature range from -150° to 500°C. Integral cooling system with dynamic covalent flow for cooling below room temperature. Sample size from 0.1 to 50 mg. Sample containers are aluminum pan type, may be used open or closed. Atmosphere

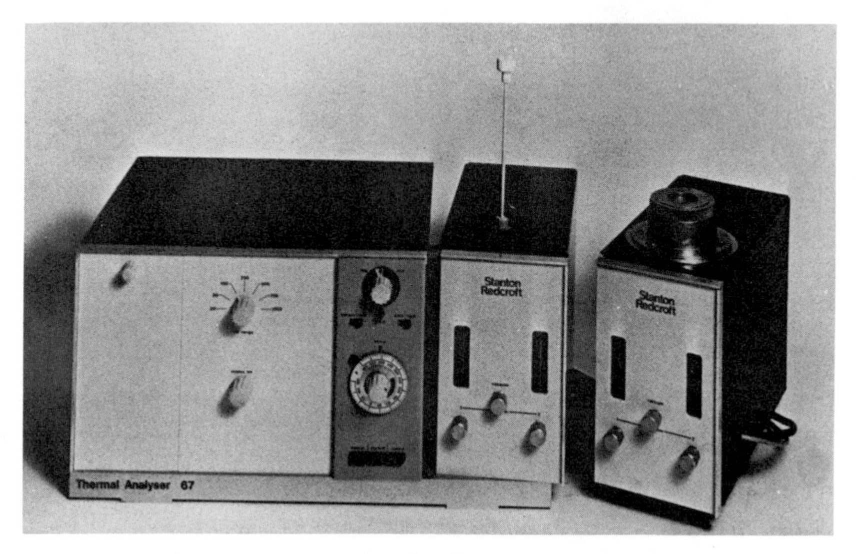

FIG. 47. Stanton Redcroft 67 system.

may be static or dynamic with air, inert, or unreactive gases. Low pressure operation to 1 x 10^{-2} Torr. Thermocouple is Chromel-Alumel with ΔT sensitivity of 0.005 to 0.10°C/mm. Visual or microscopic observation of sample.

DTA Module 2

Temperature range from room temperature to 500°C. Same specifications as for Module 1B except that the cooling system is omitted.

DTA Module 3

Temperature range from room temperature to 1000°C. Lightweight quick-cool furnace. Single stem DTA head assembly constructed of alumina with two 6 mm wells suitable for platinum or quartz crucibles. Thermocouples are Pt, Pt-13%Rh. Static or dynamic furnace atmosphere. Additional furnaces are available.

DTA Module 4

Temperature range from room temperature to 1500°C. Specifications are similar to DTA Module 3 except that the 1000°C furnace is replaced by a Pt-Rh wound furnace. The standard programmer is replaced by a linear variable programmer which has heating rates from 1.0 to 20.0°C/min. The DTA Module 4, temperature programmer, and Basic 67 Module are illustrated in Fig. 48.

It should be mentioned that all of the above systems require a two-pen strip-chart or X-Y recorder, both of which may be obtained from the manufacturer.

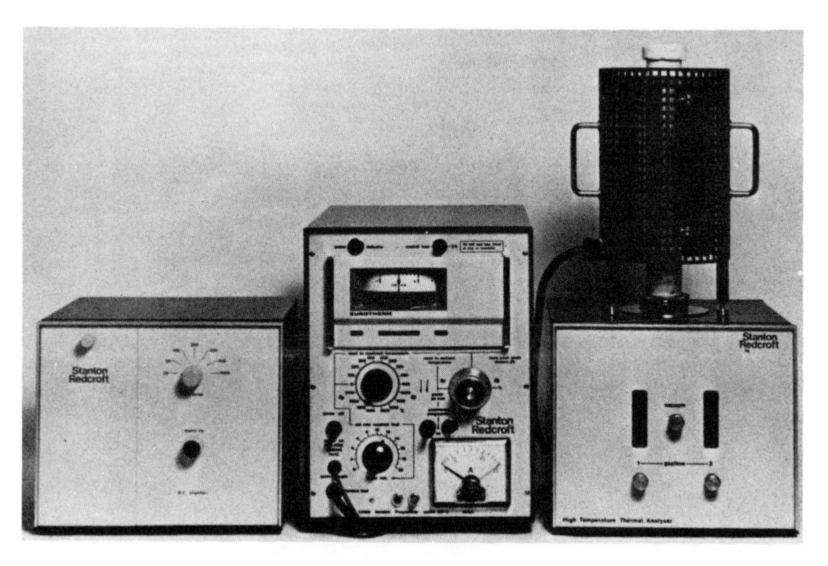

FIG. 48. Stanton Redcroft DTA Module 4, variable temperature programmer and Basic 67 Module.

TEM-PRES

The three DTA Systems, Models DT-716, DT-712, and DT-705, all operate in conjunction with the Model TA-700 programming and control console. The latter forms the basic foundation for the thermal analysis systems available from Tem-Pres. The console contains the temperature controller, the linear programmer, the dc amplifiers for the DT and T measurements, and the X-Y recorder. Provision is made in the console for acceptance of the Tem-Pres DTA, TG, DSC, dilatometer, and high-pressure DTA modules. A simple plug-in operation readily interfaces the console with any of the modules. Another added feature is that the temperature controller may be used to control or monitor any external furnace. The console, along with the TG and DTA modules, is shown in Fig. 49.

SPECIFICATIONS

Temperature range: -150° to 1600°C in 6 ranges.

Temperature program:

Operational modes: Linear temperature increase from set point - a) hold at limit, b) off at limit, c) linear down at limit; linear temperature decrease from set point - a) hold at limit, b) off at limit, c) linear at limit; isothermal; and upper and lower limits continuously variable.

Programming rates: Continuously variable; 0.1 to 75°C/min.

Programmed temperature accuracy: ±2%.

Reproducibility: ±0.5%.

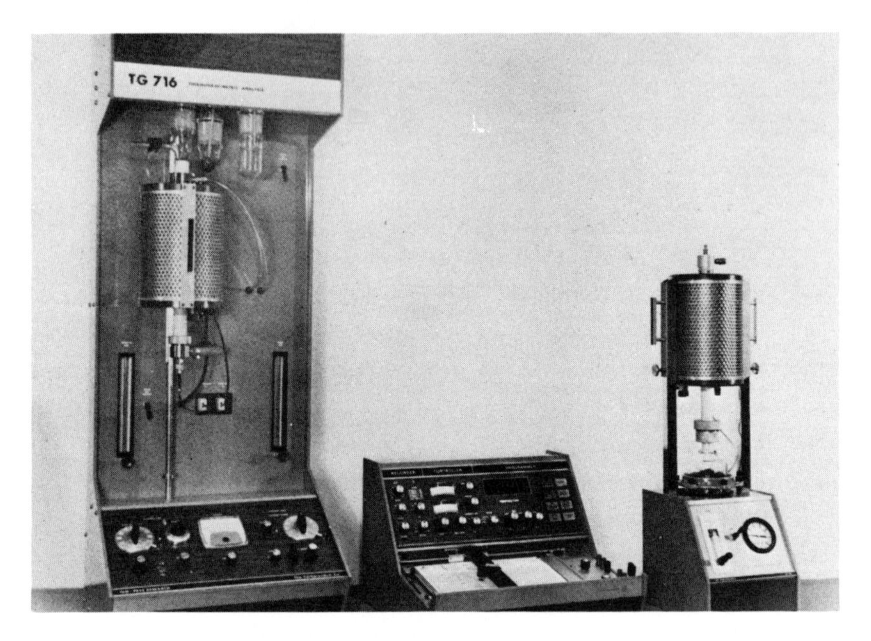

FIG. 49. Tem-Press Model TA-700 programming and control console with the TG and DTA modules.

Linearity: ±1% of full scale.

Isothermal control: ±3°C or ±25%, whichever is greater.

Temperature stability: ±0.03% per 1°C ambient, 15-40°C.

Line voltage stability: ±1°C, 110 to 130 V.

Long-term isothermal drift: ±1°C.

Maximum detectable ΔT: -.006°C Type K and T; -.025°C Type S.

ΔT sensitivity ranges: 5, 10, 20, 50, 100, 200, and 500 V/in.

Temperature ranges:

Full-scale ranges: Type K: 0-1200°C and 0-500°C. Type S: 0-1600PC, 0-1200°C, and 0-500°C. Type T: -150 to 400°C.

Range suppression: ±1 range in 6 steps, both axes.

Presentation: Simultaneous presentation of ΔT and T on 8 x 10 in. X-Y recorder and continuous digital program temperature display.

DTA Modules

The DTA systems cover the temperature range from -150° to 1600°C in five ranges using three different models. The Model DT-716 is designed for operation to 1600°C, while the other two models, the DT-712 and DT-705, operate up to 1200° and 500°C, respectively.

A typical Tem-Pres DTA system consists of the TA-700 console and one of the DT modules. The latter consists of the furnace, furnace stand, and sample holder-thermocouple assembly. By means of the console, heating rates are continuously variable from 0.1° to 75°C/min; linear programming of both heating and cooling rates is possible. Platinum sample cups are available in 50 to 200 mm^3 capacities. They are readily accessible and are held in contact with the thermojunctions by alumina support rods. Circulation of noncorrosive gases for purge or atmosphere control is possible at pressures to 45 psig. Low-pressure operation to 1×10^{-3} Torr is also a standard operating condition. A special high-pressure DTA module for operation to 45,000 psig is available.

The Model DT-716 is illustrated, along with the Model TA-700 console, in Fig. 50.

DTA SPECIFICATIONS

Temperature range: -150° to 1600°C in 5 ranges.

Temperature program:

Operational modes: Linear temperature increase from set point - a) hold at limit, b) off at limit, c) linear down at limit; linear temperature decrease from set point - a) hold

FIG. 50. Tem-Pres Model DT-716 and Model TA-700 console.

at limit, b) off at limit, c) linear up at limit; isothermal; and upper and lower limits continuously variable.

Programming rates: Continuously variable; 0.1° to 75°C/min to 1300°C; 0.1° to 30°C/min, 1300° to 1600°C.

Programmed temperature accuracy: ±2%.

Reproducibility: ±0.5%.

Linearity: ±1% of full scale.

Isothermal control: ±3°C or ±.25%, whichever is greater.

Temperature stability: ±0.03% per 1°C ambient, 15-40°C.

Line voltage stability: ±1°C, 110 to 130 V.

Long-term isothermal drift: ±1°C.

Sample size: 0.01 mg min, 100 mg max.

Sample atmosphere: Atmosphere, noncorrosive gas purge, vacuum (10^{-3} Torr), positive pressure to 45 psi.

Maximum sensitivity (ΔT): 0.1°C/in., Type K and T; 0.5°C/in., Type S.

ΔT sensitivity ranges: 5, 10, 20, 50, 100, 200, and 500 μV/in.

Temperature ranges:

Full-scale ranges: Platinel/Type K: 0-1200°C and 0-500°C. Type S: 0-1600°C, 0-1200°C, and 0-500°C. Type T: -150 to 400°C.

Range suppression: ±1 range in 6 steps, both axes.

Presentation: Simultaneous presentation of ΔT and T, on 8 x 10 in. X-Y recorder and continuous digital T display.

High Pressure DTA Module

This module consists of a pressure vessel constructed of a high-strength alloy steel, a furnace, furnace stand, and the necessary high pressure fittings and controls. Also included are noble metal sample cups and thermocouples. Control of temperature and programming is accomplished through the use of the TA-700 control console. Two models are available, the HP-DT2 with a maximum temperature rating of 950°C at 15,000 psi and the HP-DT4 for operation up to 700°C at a pressure of 45,000 psi. A diagram of the high-pressure module is shown in Fig. 51.

SPECIFICATIONS

Model	HP-DT2	HP-DT4
Pressure rating	15,000 psi at 950°C	45,000 psi at 700°C
Vessel material	Rene alloy steel body	Rene alloy steel body

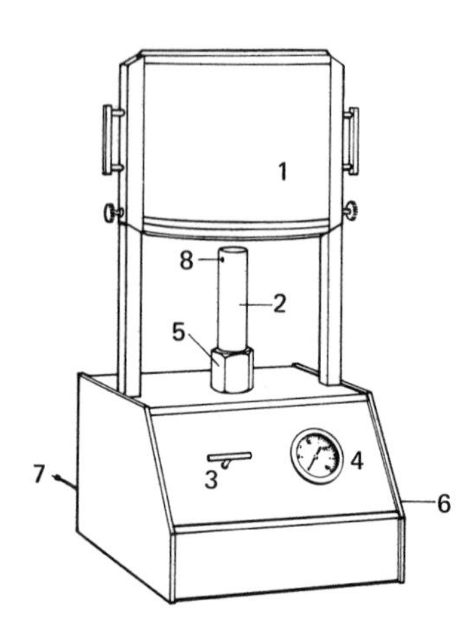

FIG. 51. Tem-Pres high pressure DTA module. 1 - Adjustable position furnace, 2 - Pressure vessel in operating position, 3 - Pressure control valve, 4 - Pressure gauge, 5 - Pressure vessel threaded closure, 6 - Pressure leads and accessories mounted within base, 7 - Control and programming leads to TA-700 Controller, 8 - Thermocouple well to permit optional vessel temperature readout.

Model	HP-DT2	HP-DT4
Vessel dimensions	1/2 ID x 8 3/4 Depth x 10 1/2 in. OA	1/2 ID x 8 3/4 Depth x 10 1/2 in. OA
Sample holders	Platinum cups	Platinum cups
Sample capacity	50 mm^3	50 mm^3
Thermocouples-Differential	Platinel 5355	Platinel 7674
Pressure fittings	1/4 in. OD high pressure	1/4 in. OD high pressure
Pressure gauge		6 in. face 50,000 psi rated
Pressure valve		1/4 in. pressure 3 way-two to pressure
Capillary assembly		24 in. L - high pressure tubing

TETRAHEDRON ASSOCIATES

Simultaneous DTA-Dielectric Cell

This cell, which is compatible with all DuPont and Mettler DTA systems, simultaneously detects the DTA and dielectric changes of a sample. It can also be used with the Tetrahedron Audrey II automatic dielectrometer or any capacitance bridge. With the former, as illustrated in Fig. 52, the apparatus will automatically record capacitance and loss tangent at any frequency within its 100- to 1000-Hz range.

Although DTA is a well-known technique, the dielectric change technique is not. The latter technique, which consists of measuring the dielectric changes of the sample with temperature, is particularly useful in monitoring and controlling curing and viscoelastic changes in polymeric materials.

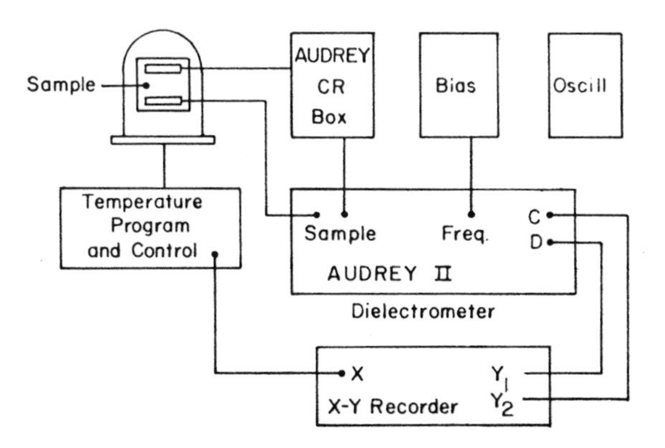

FIG. 52. DTA-dielectric change apparatus (schematic).

FIG. 53. Tetrahedron DTA-dielectric change cell.

SPECIFICATIONS

Heating rates: 0.5° to 100°C/min.

Temperature range: -200° to 600°C.

Samples: Will accept liquids, solids, films, molded disks, powders, etc.

Atmosphere: Low pressures to 10 Torr; inert or corrosive gases.

The DTA-dielectric cell is shown in Fig. 53.

THETA INDUSTRIES

Differential Thermal Analyzer, Model Labtronics 1750

The Model 1750, as shown in Fig. 54, is a low-cost, simple but versatile DTA apparatus. Three furnaces are available: -150°C to 25°C unit; 25° to 1000°C; and a high-temperature unit covering the range from 25° to 1750°C. Sample holders, of which

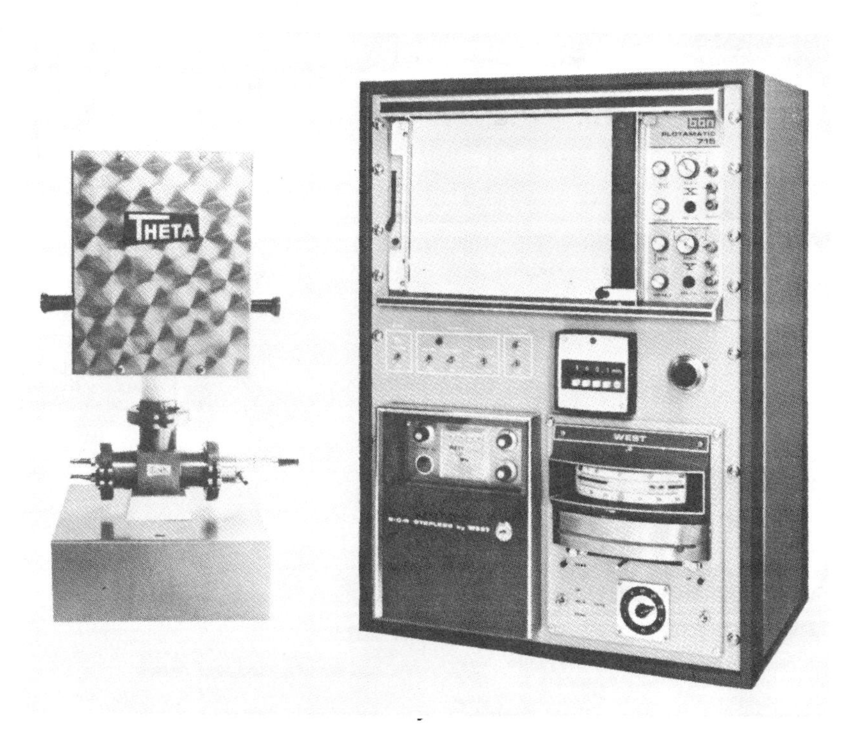

FIG. 54. Theta Model Labtronics 1750 DTA apparatus.

three different types can be chosen are the 250 μl platinum foil cylinder, a 100 μl alumina crucible, and a 10 μl copper or platinum pan. The furnace programmer controls the furnace heating rate from 0.1° to 10°C/min in 10 steps. The apparatus is modular in design with many interchangeable accessories such as thermocouple reference, metering valves, gas dryers, etc.

VOLAND CORPORATION

The Model 1100 DTA instrument is illustrated in Fig. 55. This is a low-cost instrument capable of operation over the temperature range of -100° to 1000°C in two ranges, -100 to 500°C and room temperature to 1000°C. The samples are contained in 11 x 112 mm glass or quartz ampoules which may be sealed, permitting sealed-tube studies. Sample size required is somewhat larger than for many other instruments, namely, 0.25 to 3 g. The instrument is housed in two cabinets, one of which contains the recorder. Since the recorder contains only one pen, two modes of recording are employed: 1) ΔT or sample temperature

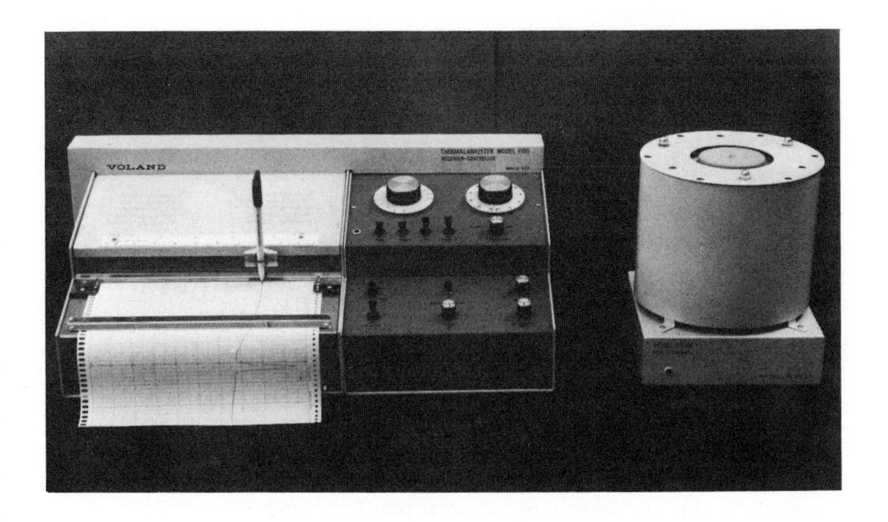

FIG. 55. Voland Model 1100 DTA instrument.

continuously, and 2) ΔT for 50 seconds and sample temperature for 10 seconds of each minute. All curves, regardless of heating rate, are displayed on 20 in. of chart.

SPECIFICATIONS

Temperature ranges: -110 to 500°C and 0 to 1000°C.

Program rates: 0.1, .167, .2, .333, .5, .6, .667, 1, 1.2, 2, 3, 5, 6, 10, and 12°C/min.

Program linearity: ±0.5% of full scale range for temperatures greater than 100°C.

Programmer modes: Linear temperature increase to either a) off at set temperature, or b) hold at set temperature.

Differential amplifier ranges: 1, 2, 5, and 10°C/in.

Zero offset: 100% on all ranges.

Presentation: 10-inch strip-chart recorder with continuous ΔT and T vs time.

Ampoules (before sealing): Volume approx. 1.5 cm^3.

Sample size: 0.25 to 3 g.

Atmosphere capability: Vacuum and static or dynamic atmospheres with inert and corrosive gases.

Power requirements: 120 V ac, 60 Hz, 750 W.

Size: Recorder-controller - 8 x 13 x 24 in. Furnace - 9 x 9 x 9 in.

Housed in heavy gauge metal, Model 1100 is supplied ready for operation with two calibration samples and one reference sample in sealed quartz ampoules (four supplied) with recessed thermocouple wells. Accessories include a) dynamic gas atmosphere and effluent gas attachment (EGA), b) ampoule sealing apparatus, c) temperature calibration kit, and d) auxiliary furnace.

Part II

THERMOBALANCES

INTRODUCTION

The technique of thermogravimetry (TG) or thermogravimetric analysis (TGA) dates back to the early 1900s. The technique consists of the continuous measurement of sample mass as a function of temperature. The sample may be heated or cooled; however, the former is nearly always the measurement employed. The mass-change curve so obtained is useful for evaluation of the sample thermal stability, the stability and composition of the intermediate compounds formed (if any), and the composition of the residue. The technique finds wide application in practically all of the fields of chemistry and other sciences.

Mass change of the sample is determined by use of a thermobalance, a schematic diagram of which is presented in Fig. 56.

A modern thermobalance consists of a recording balance, furnace and sample holder, recorder, and a temperature programmer. Sample temperature is usually measured with a thermocouple. As in the technique of DTA, furnace atmosphere must be rigorously controlled for reproducible results.

A large number of recording balances have been described, although many of the commercial instruments use the Cahn RG or RH electrobalances. Other balance mechanisms include spring and beam-deflection types as the weight transducers. The most accurate and reproducible balances appear to be the null type. Beam displacements caused by a weight change are generally detected with a light beam-shutter-photocell arrangement, although linear voltage differential transformers (LVDT) are also used.

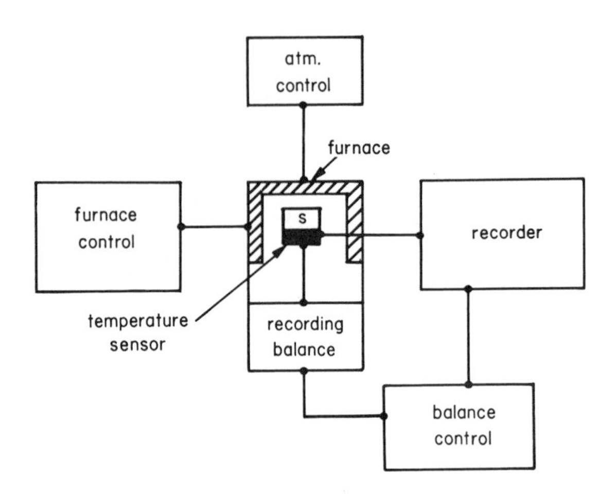

FIG. 56. Schematic diagram of a thermobalance.

The temperature range of interest determines the type of furnace chosen. The usual maximum furnace temperature is around 1000°C, although 1600° and 2000°C are also available. The choice of sample holders depends upon the maximum temperature desired; materials of construction include Inconel, aluminum, platinum, quartz, graphite, alumina, and so on. The usual configuration is a small cup, but plates, crucibles, and multiplate holders are also used.

Recording systems are generally two-channel, strip chart potentiometric types or X-Y recorders.

Commercial Thermobalances

The choice of a commercial thermobalance depends upon the nature of the application, future applications, the budget, and the laboratory space available. As in DTA instrumentation, TG modules are quite common. They may be added at any time to the basic unit, which usually contains the recording and temperature programmer. Generally, the thermobalance is needed to interpret the sample's DTA curve. It is usually impossible to identify the thermal transitions involved just from a DTA curve; complementary data from TG or other thermal techniques are needed.

There is a decided advantage in purchasing a unit which will permit simultaneous DTA-TG measurements on the same sample under identical conditions of heating rate, sample size, and furnace atmosphere. These instruments are, of course, much more expensive. They do involve a saving of time and also of interpretation. The derivative of the TG curve, differential thermogravimetry (DTG), has rather limited application unless the sample reaction kinetics are of interest. Notwithstanding many claims to the contrary, although the DTG curve will reveal minor variations in the TG curve, if a plateau is not found in the curve at this point, a quantitative calculation of the composition cannot be made. Determination of the area under the DTG peak is not an easy matter if multiple reactions are present.

The commercial thermobalances available to the investigator fall into two classifications: a) Cahn-type thermobalances, which employ one of the Cahn Electrobalances; and b) non-Cahn balance types. For the former type, the Cahn Electrobalances will be discussed first. Each manufacturer will then be discussed only in terms of the type of furnace, recording system, furnace heating rates, type of furnace atmosphere control, and any special instrumentation. Specifications for the mass-change ranges will not be repeated unless they differ drastically from the Cahn balance specifications. For non-Cahn-type thermobalances, each will be discussed in the same manner as the DTA instrumentation in Part I.

CAHN ELECTROBALANCES

RG and RH Electrobalances

By far the most popular balance is the Cahn Model RG, although many new applications are now being made of the Cahn RH balance. The main difference between them is the sample loop capacity, which is 2.5 g on the RG and 100 g on the RH model. Sensitivity is of course greater on the RG, being 5×10^{-7} μg, compared to 1×10^{-5} μg for the RH balance. The balances will record exceptionally small weight changes, down to 20 μg full scale on the recorder for the RG balance. With the zero supression system, 1% of the sample weight change can be expanded to full scale on the recorder. A wide variety of recorders may be used with the balances -- X-Y, X-T, any number of pens or pen speeds, ink or inkless, and so on. Data presentation is not limited to strip-chart or other types of recorders. Basic output is a dc voltage, which can also be applied to more sophisticated electronic data-handling equipment. Digital voltmeters are sometimes used for consecutive sample weighing and could be used, if desired, for changing samples. For very slow experiments, a microammeter may be sufficient, with total sample weight read from the balance MASS dial.

The principle of the Cahn RG balance is illustrated schematically in Fig. 57. The RH differs in that it has elastic sample bearings as well as central bearings, loop B is absent, and gold plating over the aluminum is used extensively. The balance is based on the null-balance principle, which is generally accepted as being the most accurate and reliable

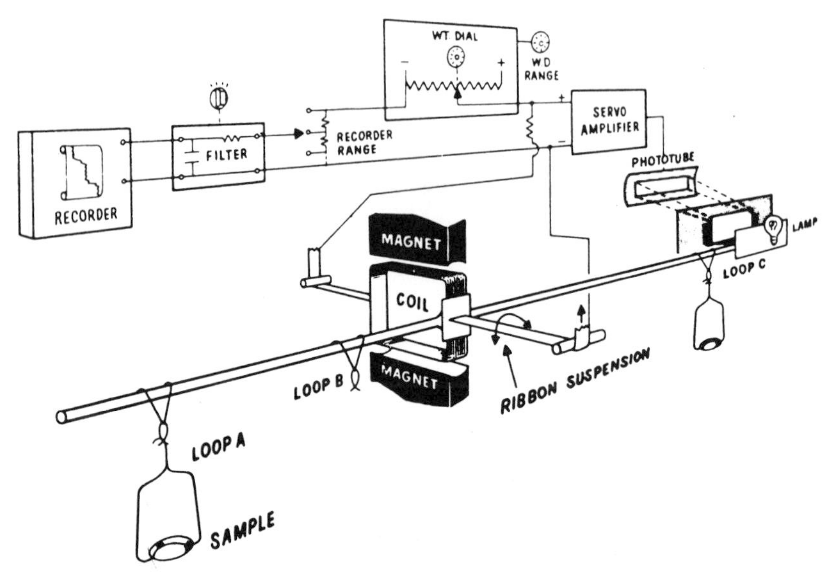

FIG. 57. Schematic diagram of Cahn RG electrobalance.

method of measurement. Changes in sample weight cause the beam to deflect momentarily. This motion changes the phototube current, which is amplified and applied to the coil attached to the beam. The coil is in a magnetic field, so current through it exerts a moment on the beam, thereby restoring it to balance. The coil current is thus an exact measure of sample weight in accordance with Ampere's law.

The elastic ribbon suspension eliminates friction and defines the axis of rotation exactly. The time constant may be set from 0.01 to 0.6 sec for the RG and 0.2 to 2 sec for the RH balance. The suspension is exceptionally rugged and difficult to damage. It is protected by stops into which it deflects when overloaded. Variations in ribbon rotational stiffness with temperature and age, and hysteresis, do not affect balance readings because of the very small angle of rotation and because readings are always made at the same angular position. The materials in the weighing mechanism are chosen for best results in high-vacuum and corrosive atmospheres.

Principal materials are aluminum, glass, and Alnico; copper and Nichrome wire are sometimes used.

The balance mechanism can be pumped down to pressures as low as 10^{-6} Torr, and even pressures of 10^{-8} Torr have been reported. A special ultrahigh-vacuum model is available for still lower dynamic and static pressures in this range. Controlled atmospheres up to 1 atm are also commonly used.

The Cahn RG and RH balances are shown in Figs. 58 and 59, respectively. Specifications are given in Table 8.

RTL Electrobalance

The recently introduced Cahn Model RTL Electrobalance is shown in Fig. 60. The balance has a top-loading configuration which can easily be converted to a thermobalance. Capacity is 10.0 g, with a sensitivity of 0.01 mg. Three full-scale ranges of weight are 0-10 mg, 0-100 mg, and 0-1000 mg, all readable to 0.1%. Analog output voltage is 0-10 V/range dc.

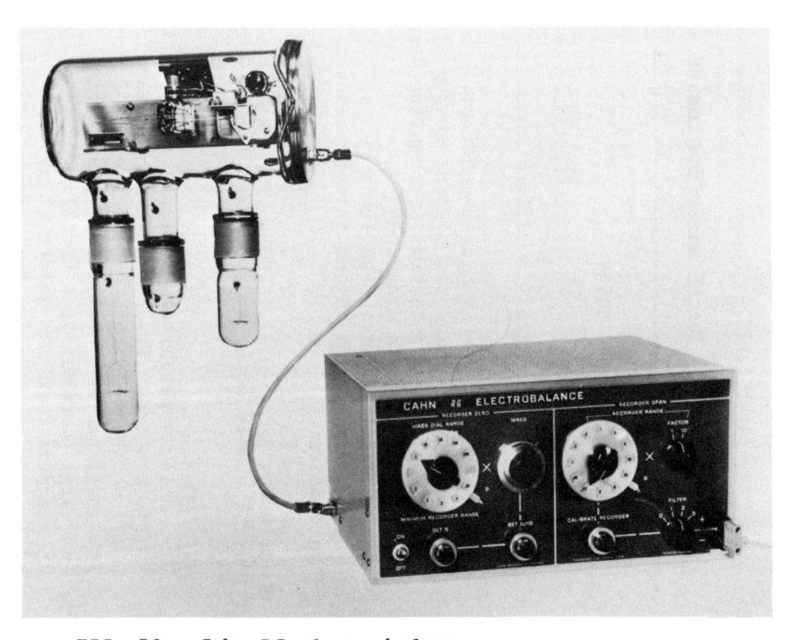

FIG. 58. Cahn RG electrobalance.

TABLE 8

Specifications for Cahn RG and RH Electrobalances

Parameter	Unit	RG Sample Suspended from loop A	RG Sample Suspended from loop B	RH Switch Position LO	RH Switch Position HI
Capacity					
Maximum load on the sample suspension: Under heavier loads the central suspension will bottom on protective, preventing damage even under severe overloads	g	1	2 1/2	100	100
Maximum weight change measurable: With extra margin beyond this at reduced precision	g	0.2	1.0	1.0	20
Sensitivity					
Weight change equal to 0.5% of a 1-mV recorder	g	10^{-7}	5×10^{-7}	10^{-6}	10^{-5}
Precision					
The smallest weight change that can be reliably detected. It depends on the total load and on the magnitude of the change					
For small samples, limited by sensitivity	g	2×10^{-7}	10^{-6}	2×10^{-6}	2×10^{-5}
For large samples, as a fraction of load		10^{-6}	5×10^{-6}	10^{-7}	10^{-7}
As a fraction of the weight change		10^{-4}	10^{-4}	10^{-4}	10^{-4}
In a given application the largest of these 3 limits determines the effective precision					

Data Presentation					
Weight changes are recorder on any 1-mV recorder					
Voltage/weight	V/g	50	10	5	0.5
The recorder span in milligrams can be varied at will while recording. The minimum span is	mg	0.020	0.100	0.200	2,000
The maximum span is	mg	200	1000	2000	20,000
13 switch-selected spans are in the sequence		1-2-4	1-2-5	1-2-4	1-2-5
The mass value for 0% on the recorder can be varied at will while recording. It is always indicated on a 10-turn dial, readable and reproducible to 0.01% of the dial range and accurate to 0.05% of the dial range (0.01% available on special order)					
The mass dial range in turn can be varied at will while recording. The mass dial ranges are	mg	0-1, 10, 20, 100 200	0-5, 50, 100, 500, 1000	0-10, 50, 100, 500, 1000	0-200, 1000, 2,000, 10,000 20,000

Both RG and RH capabilities can be combined in a single system by adding a No. 2510 RG weighing mechanism to an RH. The two weighing mechanisms may be used alternately with the same RH control unit.

Net weight: RG, 12 k, gross 15 k; RH, 25 k, gross 56 k. Control unit is 14 x 7 x 10 in. for both models. Power: RG, 33 W; RH, 35 W.

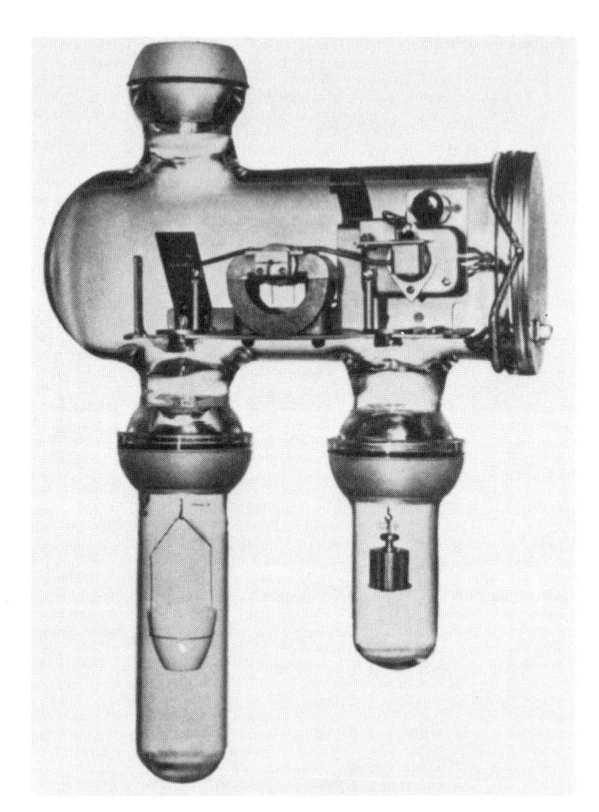

FIG. 59. Cahn RH electrobalance.

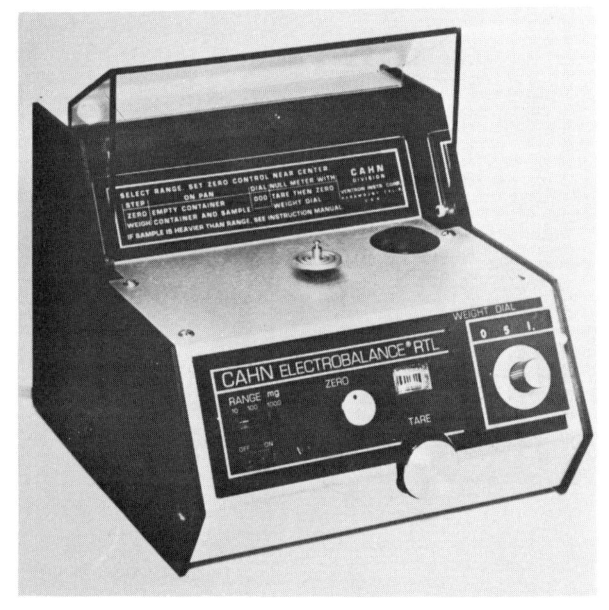

FIG. 60. Cahn Model RTL electrobalance.

CAHN DIVISION, VENTRON INSTRUMENTS

Cahn "Little Gem" TGA Kit

The "Little Gem" TGA Kit, as illustrated in Fig. 61, adapts the Cahn RG Electrobalance for TGA in air up to 650°C. The kit includes a stand for the balance, a micro furnace, Chromel-Alumel thermocouple, four disposable Pyrex hangdown tubes, Nichrome hangdown wire, and a micro sample pan. The furnace is rated at 650°C at 0.5 A and 115 V ac. Temperature of the furnace can be controlled with a variable transformer or a temperature programmer. The 5 mm in diameter sample pan will contain about 15 mg of sample. Combining the kit with a suitable recorder (X-Y or two-channel strip-chart) and an automatic temperature programmer, provides a complete TGA system.

Cahn Deluxe System

The Deluxe System, also available from Cahn, is shown in Fig. 62. This is a more advanced system, using components available from the manufacturer. The system consists of a Cahn RG balance in a vacuum bottle enclosure, a furnace, balance stand, hangdown tube and sample pans, temperature programmer, and an X-Y recorder. The temperature programmer provides furnace heating rates of 2, 4, 6, 8, 10, and 14°C/min, plus manual override.

Accessories:

The Cahn Time Derivative Computer, Mark II, produces an output voltage proportional to the rate of change of the input

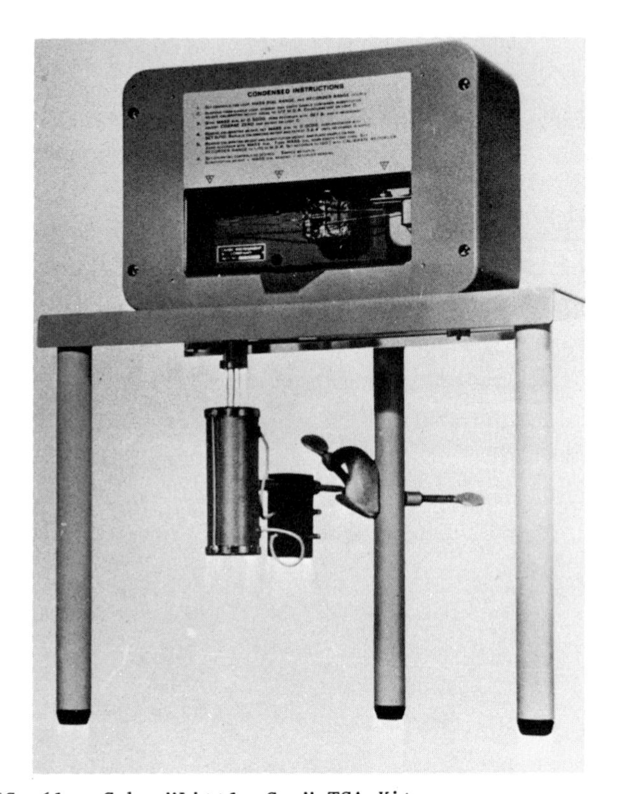

FIG. 61. Cahn "Little Gem" TGA Kit.

voltage. It was developed for use with the Cahn RG Electrobalance to compute the rate of change of sample weight. The computer is of the passive electrical analog type for maximum reliability and convenience of operation. It is designed for use with 1-mV recorders. The following full-scale ranges are obtained on a 1-mV recorder:

RG Balance: A 0-0.2, 0.4, 1, 2, 4, 10, and 20 mg/min;
B 0-1, 2, 5, 10, 20, 50, and 100 mg/min;

RH Balance: LO 0-2, 4, 10, 20, 40, 100, and 200 mg/min.
HI 0-20, 40, 100, 200, 400, and 1000 mg/min.

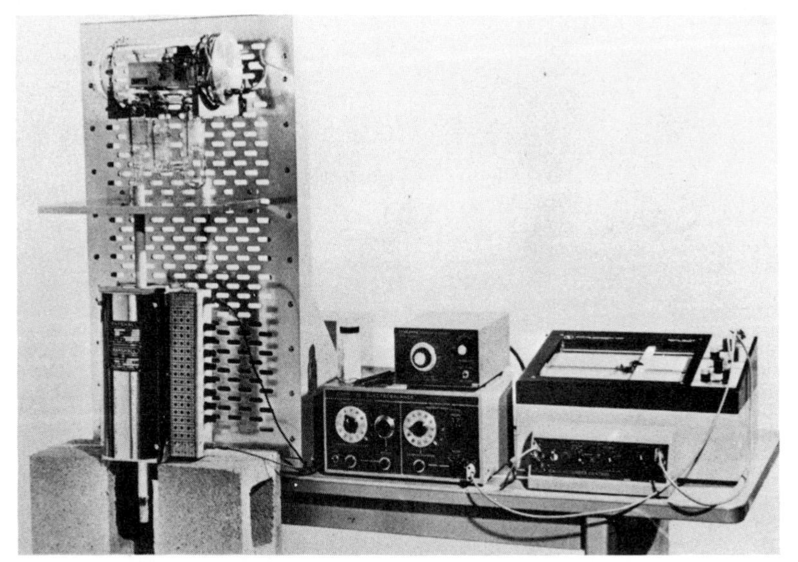

FIG. 62. Cahn Deluxe TGA system.

Exact calibration is attained by simultaneous recording of the weight and the weight derivative. The computer includes an uncalibrated zero control for setting zero rate of change to any point on the recorder chart. It requires no external power.

COLUMBIA SCIENTIFIC INDUSTRIES

Stone-Premco 1000 Series Thermobalance

The 1000 Series of thermobalances, as illustrated in Fig. 63, may be operated as a separate unit with its own temperature programmer and external recorder, or it may be used in conjunction with the Model 100 or 200 Series DTA recorder-controller (p. 17).

The furnace assembly contains a Kanthal heater element held by specially molded refractory parts. The entire furnace is of light weight construction and can be installed with ease by simply engaging the guide rods into the furnace guides and sliding the furnace up until the locking pin can be engaged. The reverse procedure removes the furnace with equal ease.

The programmer and temperature controller, P-1000 or P-1050, regulates the modes and rates of temperature increase, temperature hold, and temperature decrease. Control is achieved through comparison of a selected rate generated by the programmer with a thermocouple located in the furnace. A meter on the programmer indicates the approximate furnace temperature at any time during the run, and also serves as a set-point limit switch which the operator may set at any desired temperature point.

The unit contains all of the necessary controls and indicators for control of the gas atmosphere in the furnace chamber. Gas flow or static atmosphere may be employed.

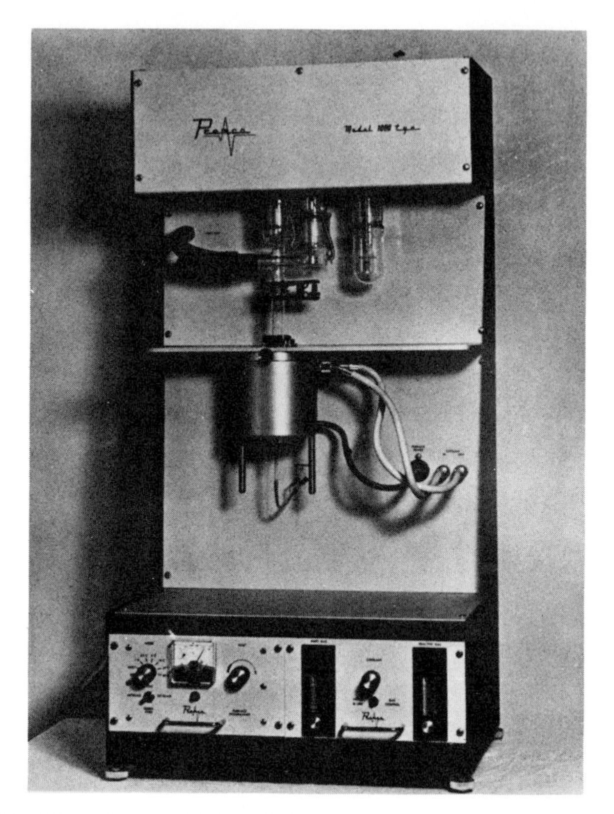

FIG. 63. Premco Model 1000 TGA system.

SPECIFICATIONS

TGA-1000-RG

Maximum capacity: 2.5 g.

Maximum weight change measurable: 1.0 g.

Maximum sensitivity: 1×10^{-6} g, recorder chart division under optimum conditions.

Temperature: Ambient to 1200°C.

Pressure: Ambient to 10^{-3} Torr.

Gas flow rate: any practical flow rate.

Sample pan size: 5 mm diameter or 9 mm diameter.

TGA-1000-RGP

Specifications are the same as TGA-1000-RG except for addition of the programmer.

Temperature program rates: 1, 2.5, 5, 10, 25, and 50°C min.

Program linearity: ±1% of full scale.

Program reproducibility: ±0.5%.

Program accuracy: ±2.5%.

Isothermal control: ±0.25% or ±3°C.

Program modes: increase, hold, decrease.

TGA-1050-RG

Specifications same as TGA-1000-RG except temperature range is ambient to 1600°C.

TGA-1050-RGP

Specifications same as TGA-1000-RGP except temperature range is ambient to 1600°C.

DELTATHERM

Deltatherm TGA System

Used in conjunction with the Basic Unit, the Deltatherm Thermobalance, as illustrated in Fig. 64, is designed for recording the weight of the sample as a function of temperature and time. The system uses the Cahn RG Electrobalance in conjunction with a furnace, which is counterbalanced on roller bearings for easy positioning of the thermal zone around the sample. The furnace is bi-filar, noninductively wound on a high alumina core. Sample pan alignment is checked by a light source and a quartz light pipe.

SPECIFICATIONS

	Model RG	Model RH
Temperature limit	1250 or 1600°C	
Recorder ranges	Loop A: 20 mg/in. to 0.002 mg/in. in 13 steps	200 mg/in. to 0.02 mg/in. in 13 steps
Maximum recorder sensitivity	Loop A: 2 μg/in.	20 μg/in.
Power	1000 W for 1250°C or 2500 W for 1600°C	
Size	24 x 24 x 71 in.; 365 lb	

Accessories include the Cahn RH 100-g capacity Electrobalance and/or the 1600°C furnace.

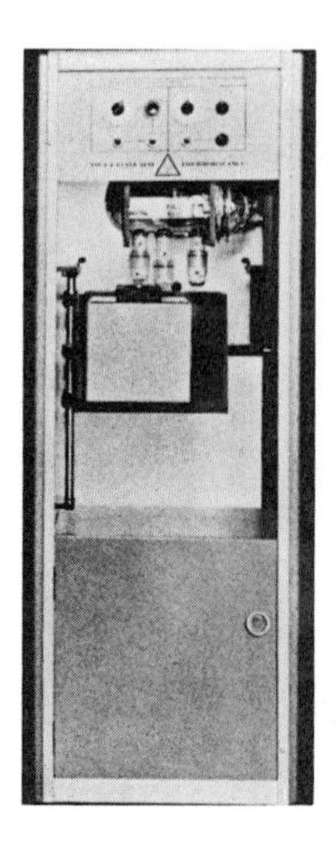

FIG. 64. Deltatherm thermobalance.

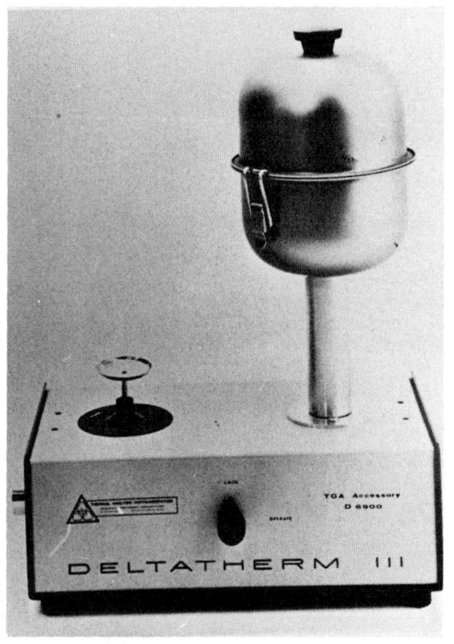

FIG. 65. Deltatherm III thermogravimetric analyzer.

Deltatherm III Thermogravimetric Analyzer

An interchangeable plug-in module to the DTA cell, the Thermogravimetric Analyzer (Fig. 65), combines an electronic null balance, and a 1200°C Kanthal-wound furnace. To achieve a sensitivity of 0.4 mg/in., the balance utilizes a Nichrome ribbon suspension.

The sample pan is easily reached by first sliding the furnace up and off the ceramic pedestal, then sliding the pedestal up and off to expose the holder. Sample weights up to 10 g can be accommodated by the balance. Pan materials are stainless steel, platinum, or quartz. Purging gases, inert atmospheres, or vacuum can be regulated through inlet ports on the back of the module.

SPECIFICATIONS

Sample system: Flat pan, above the balance, stainless steel, platinum, or quartz.

Thermocouples: Chromel-Alumel.

Sample size: Up to 10 g.

Mass amplifier ranges: 0.4 to 40.0 mg/in. in 7 steps.

Sample atmospheres: Controlled atmosphere using non-corrosive gas.

Furnace: Quick change, quick cooling, medium mass, Kanthal.

Temperature range: Ambient to 1200°C.

Size and weight: 14 x 8 x 20 in. (36 x 20 x 51 cm); 12 1/2 lb (5.7 kg).

FISHER SCIENTIFIC

Fisher TGA Accessory, Series 100

The Fisher TGA Accessory, Series 100, as shown in Fig. 66, is composed of modular components that are specifically designed to perform automatically programmed thermogravimetry in conjunction with the Cahn RG Electrobalance. In addition to the Thermocouple Reference (an electronic cold-junction compensator that eliminates the need for a reference thermocouple and ice bath), the TGA Accessory includes a linear temperature programmer, furnace, and balance stand. A 1-mV two-channel strip-chart recorder is also required.

The furnace and temperature programming unit are taken from the Model 260 differential thermoanalyzer. Sample temperature can be increased to 1200°C by eight program rates from 0.5 to 25°C/min, or the furnace may be held at a constant temperature. The TGA Accessory provides the balance cabinet, which contains a built-in jack to raise or lower the furnace. A vacuum-tight chamber holds the weighing mechanism of the balance, while the balance control cabinet is placed on the table top along with the temperature programmer. The sample (1 to 20 mg) hangs from the "A" or "B" loops of the balance, inside a quartz or Pyrex glass hangdown tube. This, in turn, fits into the tube of the DTA furnace. When the sample is suspended from the "A" loop, a 1-mV recorder will show weight changes of as little as 0.1 μg, and can measure changes of 0.2 to 1.0 μg, depending on the sample. With loop "B," it can detect a 0.5-μg weight change and measure 1 to 5 μg. Maximum changes of 200 and 100 mg, respectively, can be recorded on the two loops.

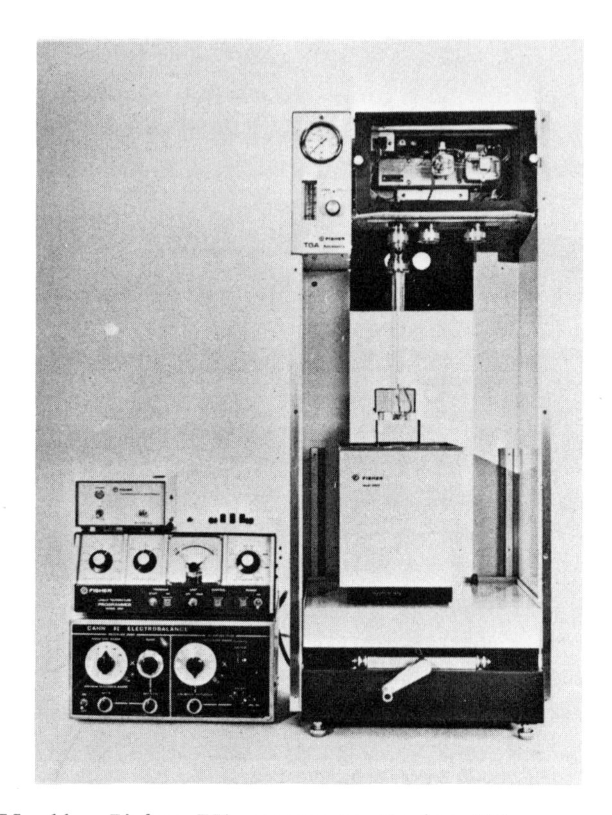

FIG. 66. Fisher TGA accessory, Series 100.

Three modes of operation are possible in the furnace chamber: (a) static environment, (b) vacuum, and (c) flowing gas stream. In the first mode the sample is heated in air, or in an inert gas at atmospheric pressure. Decomposition products diffuse away from the sample through the hangdown tube. In the vacuum mode, a single-stage mechanical pump with a free air capacity of 35 liters/min will hold the TGA system below 10 Torr. A vacuum gauge is built into the TGA cabinet. There is a connection on the back of the cabinet for a precision manometer, if a more accurate record of pressure is required. In the third mode of operation, a concentric flow tube is mounted inside the hangdown tube, enclosing the suspension wires and sample pan. The purge gas enters the inner tube, flows down past the sample, and is pumped out of the outer hangdown tube. A flowmeter is built into the cabinet. The same valve is used to control both the vacuum line and the flow of purge gas through the weighing system.

SPECIFICATIONS

Temperature range: -150 to 1200°C.

Weight (ΔW):

Maximum capacity: 1 g on Loop A; 2.5 g on Loop B.

Sensitivity: 1 μg typical of system.

Resolution: 0.01% of sample weight per inch of recorder deflection.

Calibrated accuracy: 0.01% of sample weight typical of system.

Balance ranges: 1 to 200 mg (5 steps).

Electrical tare: 100% of range.

Tare resolution: 0.1% of range.

Readout ranges: 0.02 to 200 mg (15 steps).

Temperature (T):

Approximate T ranges: 50 mV: 1300°C.

Full-scale ranges with Platinel I thermocouple: 40 mV: 1000°C; 20 mV: 500°C; 10 mV: 250°C; and 5 mV: 125°C.

Range suppression: -1 to +5 ranges (6 steps).

Precision (1/40 in. on chart): ±0.3°C (5-mV range).

Readability (1/40 in. on chart): ±0.3°C (-150 to 750°C); ±0.6°C (-158 to 1200°C).

Presentation: Simultaneous presentation of ΔW and T vs. time.

10-inch dual-channel readout (Y and Y' vs time): Simultaneous presentation of ΔW and dW/dt vs time (with accessory).

Atmospheric control:

Operating modes: Atmospheric; dynamic positive purge; dynamic reactive gas; static and dynamic vacuum.

Flow rates: 0-200 cc/min (air).

Operating pressure: μl mm to atmospheric. Vacuum of 5 μ Hg typically obtainable.

Leak rate: 1 mm in 24 hr from 1 mm absolute.

Temperature program:

Operational modes: Linear increase from isothermal -- a) hold at limit, b) off at limit; linear decrease from isothermal -- a) hold at limit, b) off at limit; isothermal.

Limits: Upper and lower limits continuously variable.

Programming rates: 0.5, 1, 2, 2.5, 5, 10, 20, and 25°C/min.

Accuracy: ±2%.

Reproducibility: ±0.5%.

Linearity: ±1% of full scale.

Isothermal control: ±0.25% or ±3°C.

Temperature stability (ambient: 15 - 40°C): ±0.03% per °C change in ambient temperature.

Line voltage stability (120 V ± 10%): ±1°C.

Turn-around time: 5 min with extra furnace.

Power requirement: 7 A on 115 V ac, 60 Hz.

Overall dimensions: 40 x 15 x 40 in. (complete system).

Shipping weight: 165 lb.

Accessories for the TGA system include a) Cahn Mark II Time Derivative Computer; b) Model 260 furnace which can be used as a second furnace, so that while one furnace is cooling, the other may be employed in the heating mode; and c) Fisher Recordall Series 300 three-pen strip-chart recorder. Use of the three-pen recorder permits the simultaneous plotting of sample mass, rate of change of mass, and the temperature.

HARROP LABORATORIES

Harrop Totalab Series TGA Module

The new Totalab Series of module-type instruments is shown in Fig. 67. The instruments are designed for applications requiring the highest possible accuracies combined with maximum versatility. They are functionally styled for attractive appearance and convenient table-top operation. Modular design permits the purchase of the complete system or one at a time, as needed.

The TGA module is designed for use to temperatures up to 1200 or 1500°C. It operates in air, vacuum, or controlled atmospheres. The furnace is mounted on a rack and pinion for easy operation and has a water-cooled stainless-steel jacket for cool operation. A gas-flow and mixing system is provided.

The Programmer-Recorder module operates the DTA, TGA, or Dilatometer modules. It features a new programming capability that allows an infinite variety of temperature program rates, cycles, and holds. An X-Y recorder records sample changes versus temperature, and a separate strip-chart recorder records temperature versus time for the actual furnace program.

SPECIFICATIONS

Temperature range: Ambient to 1200°C in Kanthal element furnace and to 1500°C in Platinum element furnace.

Furnace jacket: Water-cooled stainless steel and aluminum end plates.

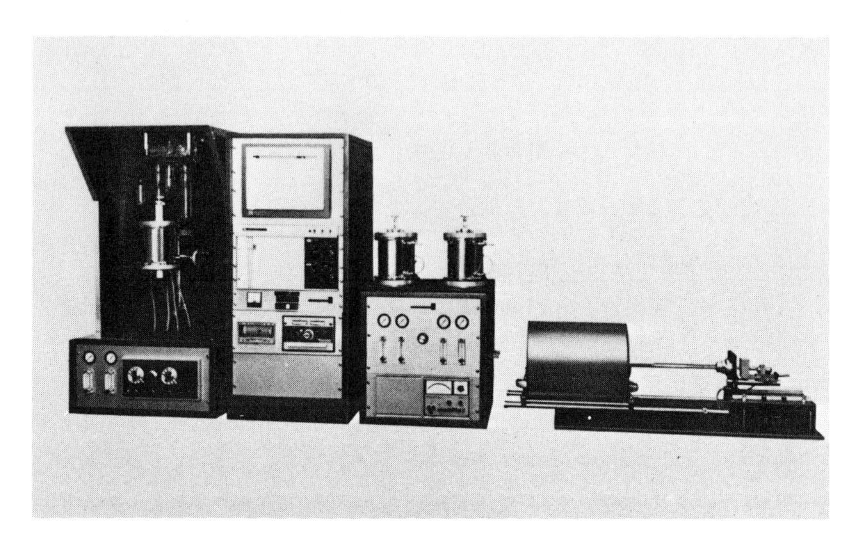

FIG. 67. Harrop Totalab thermal analysis instruments.

Thermocouples: Type "S" (Pt/Pt 10% Rh).

Maximum sample weight: 2.5 g with RG Electrobalance, 100 g with RH Electrobalance.

Maximum sensitivity: 2×10^{-5} g per chart inch with RG Electrobalance, 2×10^{-4} g per chart inch with RH Electrobalance.

Maximum weight change measurable: 1 g with RG Electrobalance, 20 g with RH Electrobalance.

Gas flow control: 0 to 0.5 LPM.

Gas pressure control: 0 to 15 psi.

Equipment also included: 1 Pt stirrup for sample crucibles. 1 Pt hangdown wire. 1 High purity recrystallized alumina flow-through tube and all components normally required for vacuum and controlled atmosphere applications. Calibrating weights and forceps.

PERKIN-ELMER CORPORATION

Model TGS-1 Thermobalance

The Perkin-Elmer TGS-1 Thermobalance couples the Cahn RG Electrobalance with a unique low-mass, high-speed internal furnace. The low-mass furnace, as shown in Fig. 68, permits programming of the temperature to 1000°C at heating rates as high as 160°C/min, with subsequent cooling to room temperature in as little as 5 min. The TGS-1 uses the control unit of the Perkin-Elmer DSC-1B (Fig. 69), but it has its own power supply and readout electronics. This allows simultaneous scanning and

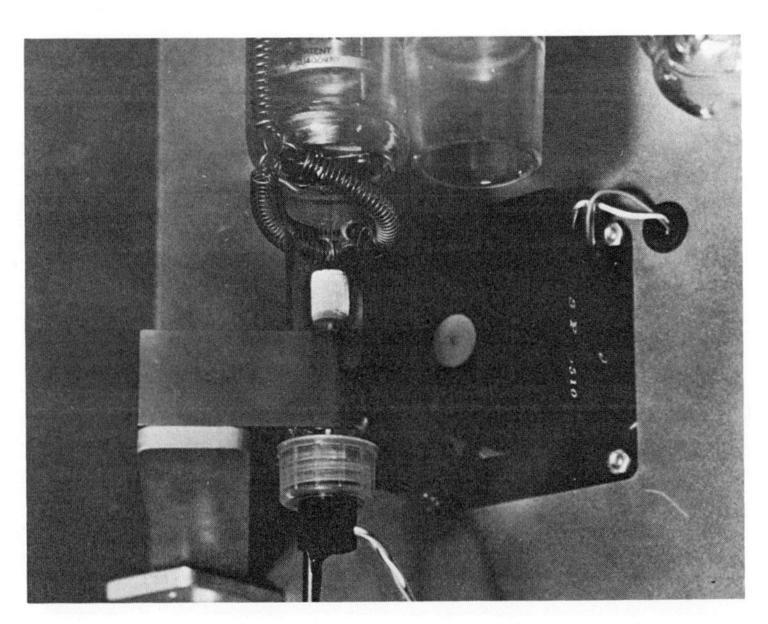

FIG. 68. Furnace of Perkin-Elmer TGS-1 thermobalance.

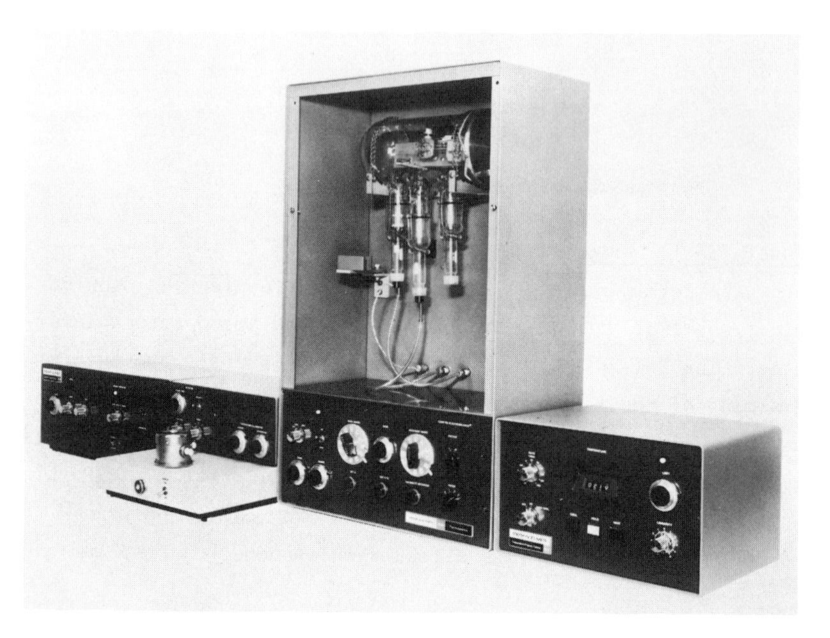

FIG. 69. Perkin-Elmer TGS-1 thermobalance connected to the DSC-1B.

recording of separate samples in the DSC and the TGS-1. The instrument may be operated in either the integral (direct weight) or derivative (DTG) mode, selectable by a front panel switch. The free space around the Pyrex hangdown tube permits easy observation of the sample and of condensable decomposition products. A permanent magnet on a pivoting mount is provided, so that the sample may be placed in a magnetic field for magnetic susceptibility measurements or for the novel Curie point temperature calibration.

SPECIFICATIONS

Sensitivity: System - 2 μg; balance - 0.1 μg.

Capacity: 1 g (including container).

Temperature range: Ambient to 1000°C.

Heating rates: 9 linear, from 0.625 to 160°C/min (0.5 to 128°C/min on 50-Hz units).

Temperature precision: ±2°C.

Temperature accuracy: ±2°C at calibration points.

Ordinate precision: 20 ppm.

Cool-down time: 5 min from 1000 to 50°C.

Weight ranges: 0.02 to 200 mg full scale (full scale = 10 in. chart width).

Operating atmosphere: Air, inert or active gas - 1 atm to 10 mm.

TEM-PRES

Two models of thermobalances, varying in the maximum operating temperature attainable, are available from Tem-Pres. The Model TGA-716 Thermogravimetric Analysis System operates up to 1600°C; the Model TGA-712 has identical specifications, except that the furnace and thermocouple are designed for a maximum operating temperature of 1200°C.

The Model TG-716, along with the Model TA-700 console, is shown in Fig. 70.

Depending upon the maximum sample sizes to be studied, the Cahn RG or RH balance may be ordered. The latter permits a 100-g sample capacity to be employed. The temperature programming and control of the furnace is supplied by the Model TA-700 console. Use of the X-Y recorders gives a convenient plot of mass change versus temperature. Controlled gas flow in the furnace chamber covers the pressure range from 1×10^{-3} Torr to 15 psig. All necessary valves, fittings, and flow meters are supplied.

SPECIFICATIONS

Temperature range: TGA-712: 0-1200°C; TGA-716: 0-1600°C.

Operational modes: Linear temperature increase from set point - (a) hold at limit, (b) off at limit, (c) linear down at limit; linear temperature decrease from set point - (a) hold at limit, (b) off at limit, (c) linear up at limit; isothermal.

Programming rates: Continuously variable 0.1 to 75°C/min.

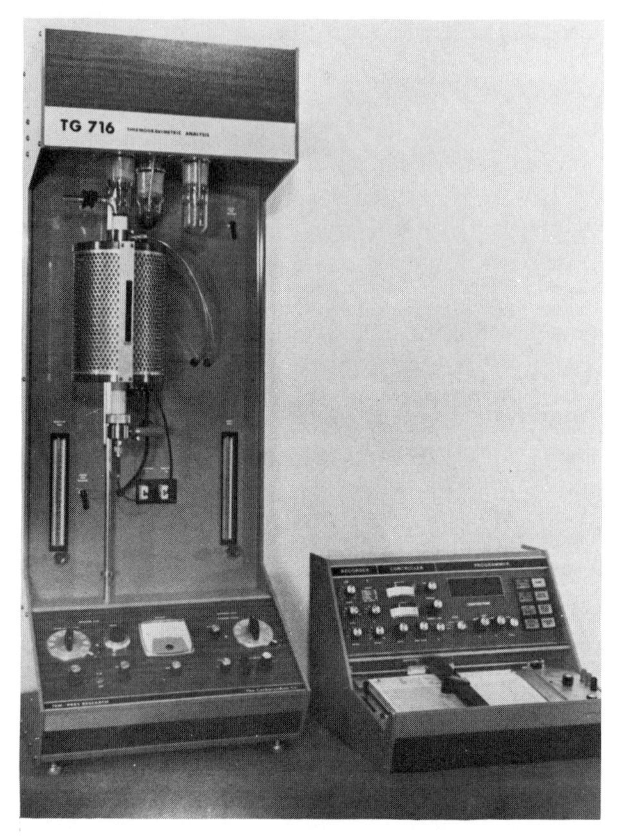

FIG. 70. Tem-Pres Model TG-716 thermobalance and Model TA-700 console.

Accuracy: ±2%.

Reproducibility: ±0.5%.

Linearity: ±1% of full scale.

Temperature stability: 0.03% per 1°C ambient 15 - 40°C.

Resolution (change in weight): 2×10^{-7} g reliably detected.

	Loop A	Loop B
Capacity:	1 g	2 1/2 g
Minimum span:	0.02 mg	0.10 mg
Sensitivity:	0.004 mg/in.	0.02 mg/in.
Maximum span:	200 mg	1000 mg
Sensitivity:	40 mg/in.	200 mg/in.

Atmospheric control: Vacuum to 10^{-6} Torr to 100°C, 10^{-3} Torr to 1600°C. Inert and reactive gas capability to 15 psi. Control and programming of TGA is supplied by Tem-Pres TA-700 control console. X-Y recorder for indicating weight change and temperature is an integral part of the console. TA-700 Control also accepts Tem-Pres plug-in modules for DTA-DSC and TD analysis units.

Power requirement: 15 A - 115 V ac, 60 Hz.

Dimensions and weight: 18 x 16 x 47 in., 95 lb.

A.D.A.M.E.L.

Chevenard Thermobalance

The Chevenard thermobalance was one of the first commercial thermobalances. It was developed for metallurgical applications by P. Chevenard in 1939 and applied to the study of analytical precipitates by C. Duval and his coworkers in the early 1950s. It was, at one time, the "standard" thermobalance for thermogravimetric investigations.

The principle and mechanism of the Chevenard thermobalance are shown in Fig. 71. The sample is placed at the end of support rod T, which is suspended at one arm (F) of a sensitive balance. The other end of the arm contains a balance counterweight, C_p. The arm can rotate in a vertical plane by means of two suspension wires, r_1 and r_2, which are made of a special alloy. Support rod T is suspended from the arm by a third suspension, r_3, of the same type as r_1 and r_2. Any parasitic vibrations of the assembly are removed by damping device A. The sensitivity of the balance is adjusted by moving counterweight C_s.

The sample is heated or cooled by an automatically controlled furnace. Support rod T can be positioned parallel to the long axis of the furnace chamber. Temperature of the furnace and/or sample is measured by a Pt, Pt-10%Rh thermocouple, whose hot junction, C, is positioned near the sample container.

For recording the weight-change curve, a small mirror, M, is attached to one end of balance arm F. A small beam of light from source S is reflected from this mirror and converges to

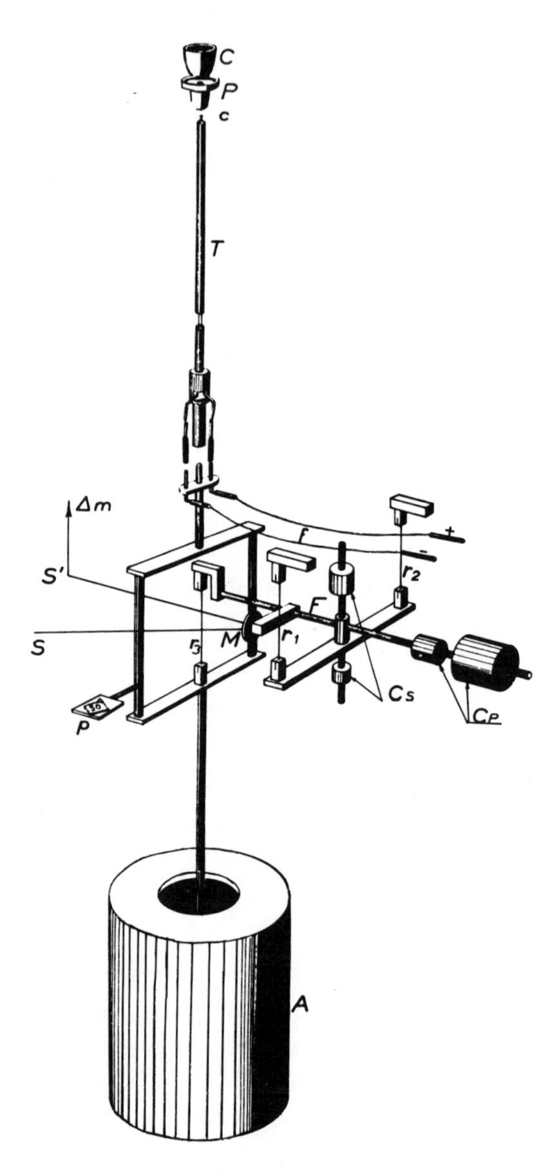

FIG. 71. Schematic illustration of the Chevenard thermobalance weighing mechanism.

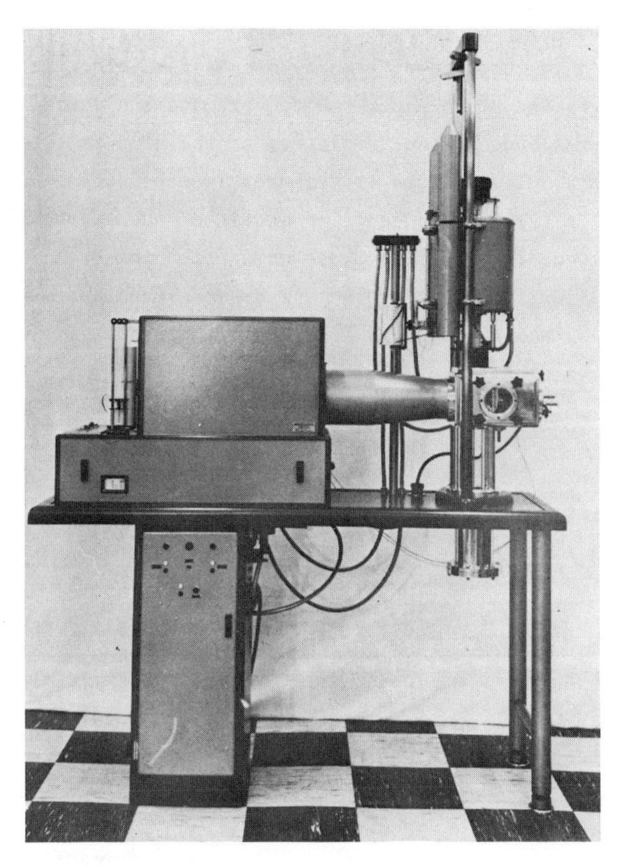

FIG. 72. A.D.A.M.E.L. Chevenard thermobalance TH-59, Model C.

form a spot, S'. Any change in the sample mass, m, results in a change in position of the arm F, and hence movement of light spot S' in vertical direction. Depending upon the particular model of thermobalance, the displacement of this spot will either be recorded directly on photographic paper or else graphically on chart paper. In the case of the latter, the pen that traces the curve is connected to a photocell designed to follow all displacements of the light spot reflected by the mirror.

In the simpler models, the change in sample mass is recorded as a function of time by means of an electric motor which moves the graph paper. Speed of rotation normally provided corresponds to one revolution of the drum in either three or 24 hours. Temperature is measured independently and written manually on the curve. In later models of the thermobalance, the thermocouple emf is recorded on a potentiometric recorder.

The specifications of the various thermobalances are given in Table 9, while the TH-59 Model C thermobalance is shown in Fig. 72.

TABLE 9

Specifications of Chevenard Thermobalances

	TH 46 Model 1	TH 46 Model 2	TH 59 Model 1	TH 59 Model 2
Range	10 g	10 g	20 g	20 g
Sensitivity	0.5 to 2 mm/mg		0.5 to 5 mm/hg	
Mode of recording	Photographic		Spot-follower recorder	
Max. temp.	1050°C	1050°C	1050°C or 1500°C	
Approximate weight	65 kg	95 kg	1050° furnace 200 kg 1500° furnace 215 kg	
Size	1600 x 400 x 1250 mm	1760 x 400 x 1350 mm	1651 x 508 x 2358 mm	

AINSWORTH

Type 15 Vacuum Balance

Although a thermobalance is not available from this company, their balances are used as components in thermobalance systems. The balance most commonly used for this purpose is the Type 15 Vacuum Balance, illustrated in Fig. 73. This balance is based on an electronic null-restoring system. Beam motions are detected with an audiofrequency variable inductance transducer whose signals are amplified, rectified, and applied to special coils attached to each beam end. The coils move in a permanent-magnetic field, the electromagnetic forces thus

FIG. 73. Ainsworth Type-15 vacuum balance.

developed restore the beam to its initial position, and the coil current becomes a measure for the weight on the balance pan.

SPECIFICATIONS

Capacity: 200 g.

Sensitivity: Can be set by switch to give output voltage of 100 mV for 10 mg, 100 mg, 1 g, 10 g, 25 g.

Reproducibility: S = ±0.1%.

Accuracy: Linearity of ±0.1%.

Recording ranges: 5 ranges.

Response time: Less than 0.5 sec.

Maximum ambient temperature rise: 125°C.

Typical vacuum: 10^{-6} Torr.

AMERICAN INSTRUMENT COMPANY

Aminco Modular Thermo-Grav

The Aminco Modular Thermo-Grav combines in one unit the advantages of vacuum thermogravimetry, differential thermal analysis, and effluent gas analysis. In addition to its greater flexibility, the instrument requires less floor space and affords savings over the purchase of separate DTA or TGA units. The instrument consists of a completely integrated DTA/TGA apparatus (Effluent Gas Analyzer may be ordered as an accessory), in which the temperature programmer and X-Y recording system are linearized for Chromel-Alumel thermocouples and are compatible for either the DTA furnace and amplifier or the TGA furnace and weighing system. Change of mode is accomplished by a selector switch and rotation of furnaces within the base assembly. The Modular Thermo-Grav is illustrated in Fig. 74.

Three options are available: a) the Basic TGA Unit; b) the Modular DTA Unit; or c) the complete DTA-TGA-EGD Unit. All are illustrated in Fig. 74.

The thermobalance, in contrast to electronic null-type instruments, uses a spring to measure changes of sample weight. The sample is placed in a crucible supported by the precision spring suspension system which hangs freely within a Pyrex-quartz glass enclosure. The coil windings of a displacement transducer (linear voltage differential transformer) are mounted outside the glassware enclosure, concentric with a soft-iron armature fastened to the suspension system. Any change in sample weight during heating causes a corresponding

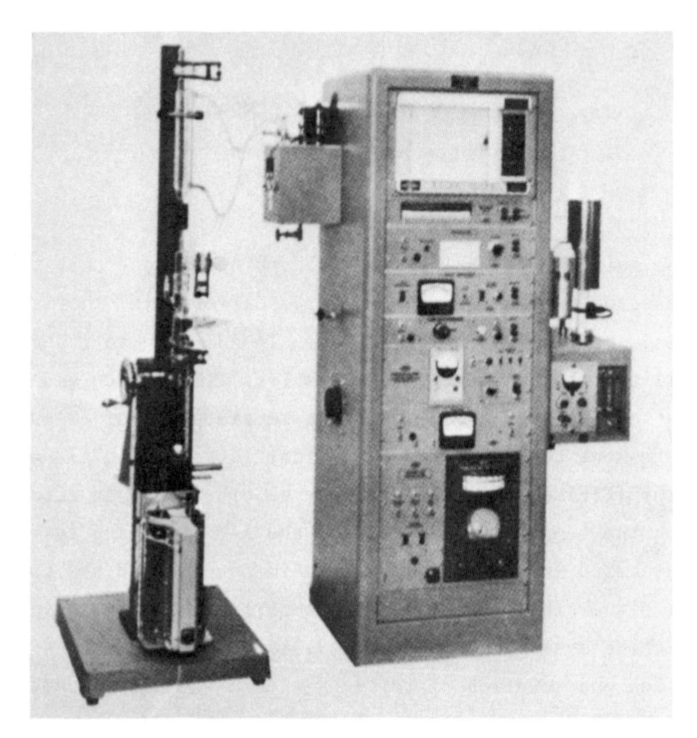

FIG. 74. Aminco modular thermograv.

vertical displacement of the suspension and generates an ac voltage proportional to it. The demodulator converts the signal to dc so that it can be recorded on the X-Y recorder. The recorder weight-change axis is calibrated by adding or removing known calibration weights to or from the suspension system, simulating the anticipated weight change. The calibration signal voltage can be adjusted at the demodulator to obtain the most convenient pen deflection in milligrams per inch.

The sample chamber can be flushed continuously with gas during the run. The gas supply is connected to the vacuum inlet of the glassware enclosure and flows down into the sample chamber, past the sample, and out through the small holes at the tip of the gas well. This type of flow system practically eliminates recondensation of vapors on the suspension system and allows cold trapping of condensable gases as they leave the sample chamber.

SPECIFICATIONS

TGA Mode

Data presentation: Weight change vs sample temperature; weight change vs elapsed time.

Sample capacity: Any sample weight up to 10 g; up to 230 g optional.

Recording range: 0 to 200 mg of weight change; 0 to 5 g optional.

Sensitivity: 1 in. of pen deflection for 5 mg of weight change; continuously variable adjustment to lower sensitivities.

Accuracy: ±0.5%.

Temperature range: Ambient to 1050°C.

Heating rates: 1 1/2, 3, 4, 4 1/2, 6, 8, 9, 12, 16, and 24°C/min, using gear changes and pre-cut template (supplied).

Furnace: 1050°C, bifilar wound, equipped with thermal fuse and Chromel-Alumel thermocouple sheathed in stainless steel.

Sample thermocouple: Chromel-Alumel in alumina double-bore beads.

Vacuum: 0.1 Torr, readily attainable with mechanical vacuum pump system supplied; glassware can be evacuated to 10^{-6} Torr or less with 2-in. diameter, diffusion pump; thermocouple-type vacuum gauge tube and meter calibrated 0-1000 μ Hg.

Gas flows: All necessary adapters for connecting the Pyrex quartz enclosure to an external gas supply.

Static atmosphere: System can be purged and sealed, or left open to air during test.

Specifications on the DTA mode have been presented in Part I under the Aminco Thermoanalyzer.

EGD System (Optional). The Effluent Gas Detector is a thermal conductivity bridge having one thermistor in each leg. The thermistors are mounted one each in the sample and reference gas streams. The bridge is then balanced with a given carrier gas flowing through the system. As gaseous products evolve from the sample during a run, they are swept past the sample thermistor, causing a bridge unbalance voltage. This voltage is then recorded on a separate recorder simultaneously as the DTA curve is recorded on the X-Y recorder.

SPECIFICATIONS

EGD Mode

Detection: By thermal conductivity bridge containing two thermistors, one each in the sample and reference gas streams.

Power supply: Self-contained, solid state, regulated with Zener diodes; adjustable from 10-22 V.

Carrier gas: Any gas may be used; internal calibration circuit indicates the proper voltage to use for maximum sensitivity.

Sensitivity: 10-position switch for reproducible step-attenuation of bridge unbalance voltage from 1:1 to 1:1000.

Recording: For simultaneous recording of DTA and EGD, a 1620-832 strip-chart recorder in addition to the 1620-831 X-Y recorder is required; or a 1620-837 two-pen (X-Y,Y') recorder in place of both.

DU PONT INSTRUMENTS

The Du Pont 951 Thermogravimetric Analyzer is one of a number of individual modules which plug into the Du Pont 990 Thermal Analyzer. This unit utilizes the temperature programmer and X-Y recorder of the 990 system.

The thermobalance is illustrated in Fig. 75. The balance mechanism contains a null-balancing, taut-band electric meter movement with an optically actuated servoloop. The sample is placed in a sample container which is suspended directly on the balance beam. In normal operation, the furnace temperature-sensing thermocouple is positioned within 1 mm of the sample, and hence, indicates very close to the true sample temperature.

The mass loss or gain of the sample is plotted on the X-Y recorder, or mass on the Y axis and temperature (or time) on the X axis. Mass changes as small as 2 μg may be detected on the recorder. For large samples, electrical compensation up to 100 mg may be employed with mechanical taring above 100 mg.

The beam-in-furnace design and position of the sample container permits axial flushing of the furnace tube with various gases. The sample presents minimum cross section to the gas flow, which results in negligible torque perpendicular to the beam. This permits rapid change of sample environment with a minimum of turbulence and noise. TG curves may be obtained under vacuum, atmospheric pressure, or positive flushing pressure.

The furnace can be programmed to 1200°C at a rate continuously variable from 0.5 to 30°C/min. It will attain a tempera-

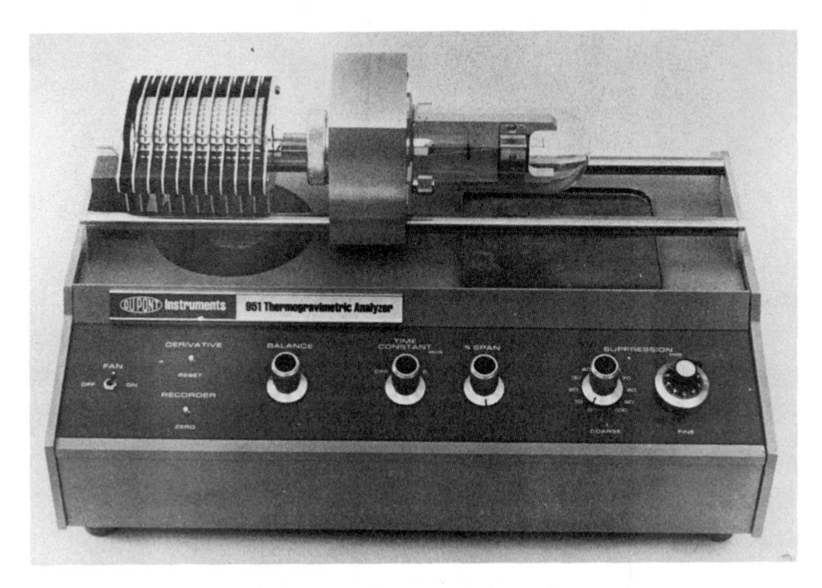

FIG. 75. Du Pont 951 thermogravimetric analyzer.

ture of 1200°C from ambient room temperature in a minimum time of 15 min. The furnace is of the plug-in-modular type with a low thermal mass.

A derivative of the mass-loss curve can also be obtained if a two-pen X-Y_1Y_2 recorder is employed. The sensitivity of the derivative circuit is 5 μg/min. By the use of scale expansion on the Y axis, as small a change as 0.5% of the sample mass can be displayed full scale.

SPECIFICATIONS

Capacity: 1000 mg (including boat).

Mass ranges: 1, 2, 5, 10, 20, 50, and 100 mg full scale (5 in.).

Temperature range: To 1200°C.

Suppression: 100.00 mg (stepped and continuously variable).

Suppression accuracy: ±0.02 mg.

Sensitivity: 0.2% of full scale (0.002 mg ultimate).

Precision: ±0.2% of full scale (±0.002 mg ultimate).

Accuracy: ±0.2% of full scale (±0.002 mg ultimate).

Noise band: 0.008 mg (zero time constant); 0.002 mg (1-sec time constant). Unaffected below gas flow rates of 2.5 l/min.

Time constant: 0, 1, 2, 5, 10, and 20 sec.

X-axis time base: 0.5% linearity; variable, 1 to 200 min. full scale.

Dimensions and weight: 9 x 12 x 19 in.; 20 lb.

LINSEIS

L 81 Thermobalance

This thermobalance is an updated model of the L-72 Series thermobalances, and many of the sample holders and accessories are similar.

The L 81 thermobalance is capable of either horizontal (L 81/2) or vertical (L 81/3) operation; both modes of operation have certain advantages and disadvantages. For example, with vertical operation, the sensitivity is reduced to approximately one-half. Also, with this mode of operation, the maximum temperature can be as high as 2200°C since the balance beam is not heated. The vertical design ensures more favorable centering of the sample container, so that with DTA higher sensitivities can be employed. The horizontal position is genrally used only up to 1100°C for TG measurements in atmospheres requiring rather rapid gas flows. Both inert and corrosive gases can be passed through the sample chamber. A faster heating rate may also be employed. The horizontal version of the thermobalance is shown in Fig. 76.

The thermobalance is capable of operation in inert and corrosive atmospheres at atmospheric pressure and in low pressures down to 1×10^{-5} Torr. All parts of the weighing system are constructed of stainless steel, Teflon, gold, or quartz. The balance can be converted for DTA measurements by the addition of the sample and reference chambers and the appropriate electronics equipment.

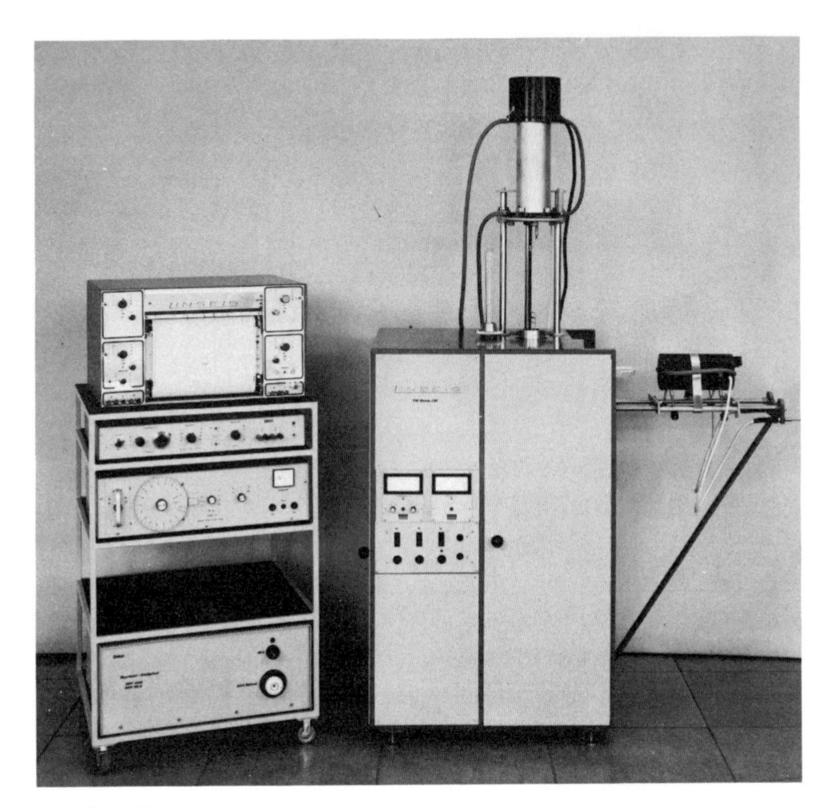

FIG. 76. Linseis L 81/2 thermobalance.

The various furnaces, covering the temperature range from -150 to 2200°C, are illustrated in Fig. 77. These furnaces are for use in the vertical mode only. The horizontal furnace has a temperature range from 200 to 1250°C.

SPECIFICATIONS

Measuring range (as selected):

Horizontal	2	20	200	2,000 mg
Vertical	5	50	500	5,000 mg

L 81/260

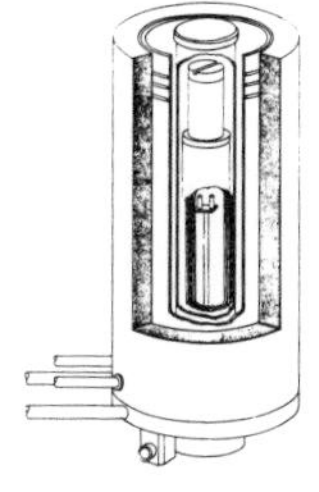

Temperature range	—150 to +350 °C
Heater	bifilar Kanthal wire
Control resistance	Pt 100 thermometer
Internal diameter	18 mm
Const. temp. zone	20 mm approx.
Coolant	liquid nitrogen
Loading	0,3 kVA

L 81/290

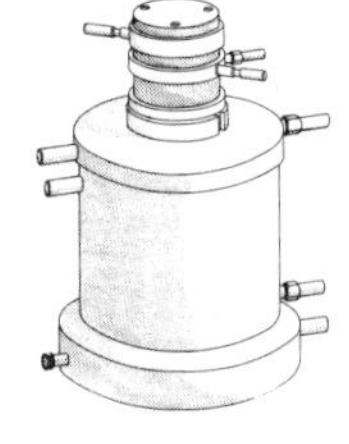

Temperature range	20—2200 °C
Heater	graphite tube
Control thermocouple	W/Re 5% — W/Re 26%
Internal diameter	15 mm
Const. temp. zone	20 mm approx.
Cooling time	30 min.

L 81/230

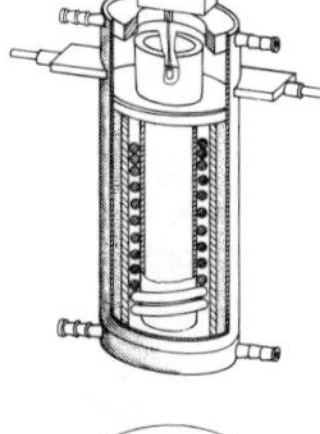

Temperature range	20—1250 °C
Heater	bifilar Kanthal wire
Control thermocouple	Pt-Pt/Rh
Internal diameter	36 mm
Const. temp. zone	30 mm
Loading	1,3 kVA
Cooling time	1 hour

L 81/240

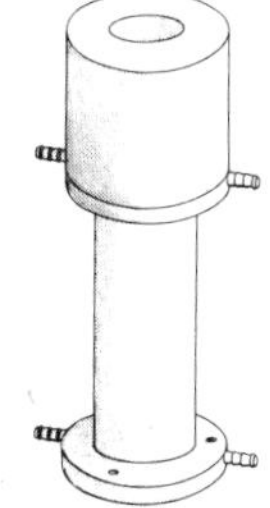

Temperature range	20—1600 °C
Heater	bifilar SiC tube
Control thermocouple	Pt-Pt/Rh (EL 18)
Internal diameter	40 mm
Const. temp. zone	25 mm approx.
Loading	3 kVA
Cooling time	2 hours

FIG. 77. Furnaces for vertical arrangement of the Linseis L 81 thermobalance.

Maximum electrical weight compensation:

Horizontal	250	2,200	4,000	4,000 mg
Vertical	750	6,600	12,000	12,000 mg

Maximum sample weight:

Horizontal	4,000	10,000	20,000	20,000 mg
Vertical	10,000	25,000	50,000	50,000 mg

Sensitivity:

Horizontal	0.002	0.02	0.2	2 mg
Vertical	0.005	0.05	0.5	5 mg

Reproducibility: Horizontal/vertical ±0.05%, 0.1%.

Temperature drift of zero: 0.01 mg/°C.

Temperature drift with 100 x weight suppression: 0.04 mg/°C.

Tare range: ±5 measuring ranges.

Weight suppression: Coarse with 1 x, 10 x, and 100 x; fine with 10-turn potentiometer and Duodial knob.

Accuracy: 0.2%.

Linearity: 0.1%.

Resolution: 0 004.

Stabilization factor: 10000.

Amplifier reference voltage: 10 mV.

Output impedance: 10 Ω.

Response time of system: Less than 1 sec.

Time constant of output filter (selected by switch): 0.5 sec/5.5 sec.

Netzsch Thermobalance 409

This thermobalance, the balance and furnace of which are illustrated in Figs. 78(a) and 78(b) respectively, consists of a three-knife-edged balance equipped with air-damping. The balance has a sensitivity of 0.1 mg. The sample container has a diameter of 30mm. A suspension chain on the opposite side of the balance beam permits taring of the sample up to 2 g. The weight variations during the run are compensated for by a second

FIG. 78(a). Netzsch Thermobalance 409: interior of balance.

FIG. 78(b). Netzsch Thermobalance 409: furnace and sample holder assembly.

chain whose effective length is automatically adjusted by a servomotor. Control of the servomotor is effected by a photocell circuit which senses the unbalance of the beam. Automatic recording of weight variations up to 100mg is permitted.

Either a tube furnace with a Kanthal heater element for a maximum temperature of 1320°C or a silicon carbide heater tube for a maximum temperature of 1550°C is used as the furnace. The furnace can be moved up and down via a guide rail. The weight of the furnace is compensated by counter weights, which are installed inside the supporting column. In order to avoid instability caused by air convection currents and heat radiation, the balance is equipped with radiation and convection shields.

Temperature Control Unit 406 is used to control the furnace temperature. The heating rate is adjustable in ten steps and equipped with a limit switch for four ranges.

The weight-change curve is recorded on a two-pen recorder. One pen records the weight change, the other the temperature, both as a function of time. Chart width for the recording is either 120 or 250 mm. Three different chart speeds are provided.

SPECIFICATIONS

Temperature range: 20 to 1320°C; 20 to 1550°C.
Accuracy: ±0.5 mg.
Automatic range: ±200 mg.
Maximum capacity 50 g.
Temperature control: 0.5%.
Power required: 1.5 kW or 3.5 kW (depending on furnace).

Netzsch Vacuum Thermobalance 419

This thermobalance is a sophisticated version of the Thermobalance 409 and can be used under high vacuum conditions as well as atmospheric pressure. Maximum sample mass is 10 g with a range of automatic weight compensation of 200 mg or ±100 mg.

SARTORIUS

Thermobalances

There are three models of the Sartorius thermobalances: the Gravimat Series 4300, the Thermo-Gravimat Model 4304, and the Thermomat S Model 4204. The Gravimat model is used mainly for the measurement of the specific surface and pore size distribution, sorption capacity, and so on, of solids. By the attachment of a furnace, it can be used for thermogravimetry studies in the temperature range from 25 to 600°C (standard) or to 2200°C (on request).

The Thermo-Gravimat is a modification of the Gravimat and is used for thermogravimetric investigations in the temperature range from -220 to 2200°C and at pressures from 5×10^{-4} to 800 Torr. Also, simultaneous TG, DTA, and DTG data can be recorded, and a mass spectrometer can be attached to the system for evolved gas analysis. The main advantage of this apparatus is that measurements of specific surface and pore size distribution can be carried out before and after the TG experiment.

For TG studies in very corrosive atmospheres, the Thermomat S instrument may be employed. It uses a magnetic suspension balance in which the weighing system and sample chamber are hermetically sealed from each other. The apparatus is illustrated in Fig. 79.

The apparatus consists of an electronic microbalance, a vacuum pump, furnace, temperature programmer, and various electronic circuitry. It can simultaneously record the TG, DTG, and DTA curves of a sample under various controlled pyrolysis conditions.

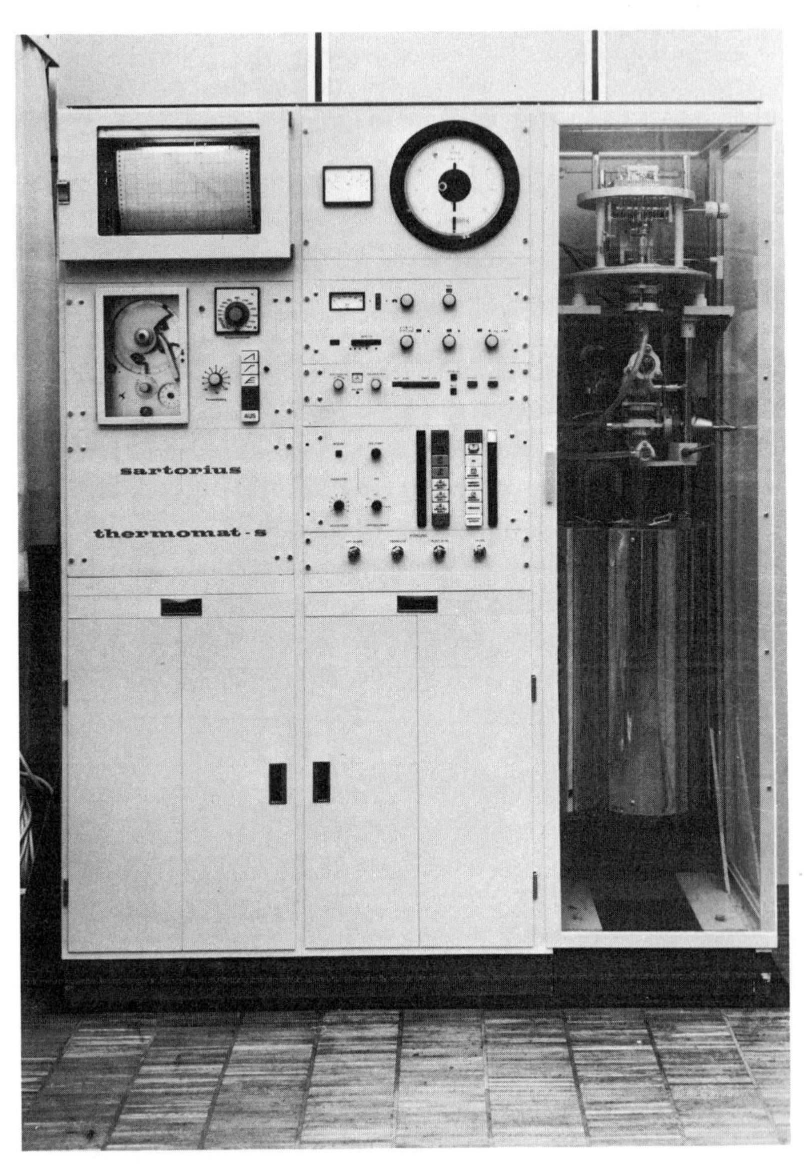

FIG. 79. Sartorius Thermomat S Model 4204.

As indicated in Fig. 80, there is no mechanical connection between the balance and the sample chamber. This coupling is made by electronically controlled suspension magnets. Mechanical weights, electric weight compensation, and electric tare compensation permit weighings up to a maximum load of 30 g with a smallest detectable weight change of 10 μg.

The pump system consists of a diffusion pump combined with a water jet pump as a backing pump and a mercury diffusion pump. Two cold traps prevent vapor diffusion into the weighing chamber.

A number of interchangeable furnaces are available for use in the temperature range of -196 to 2200°C.

SPECIFICATIONS

Balance

Maximum load: 40 g.

Electrical weighting range: 10 μg to 20 g.

Electrical weight compensation: 1 mg to 19.9 mg.

Measuring ranges: 1 mg to 10 g full scale in 9 ranges.

Maximum sensitivity: 10 μg.

Vacuum System

Water jet pump capacity: 3601/hr.

End vacuum: 17 Torr.

Mercury diffusion pump capacity: 6 liters/sec.

End vacuum: 10^{-4} Torr.

Operating vacuum: 5×10^{-6} Torr.

Two liquid N_2 cold traps.

Heating and Cooling Unit

Temperature range: -196 to 1150°C.

Tube furnaces to choice.

Linear programmed heating up rates: 0.5 to 30°C.

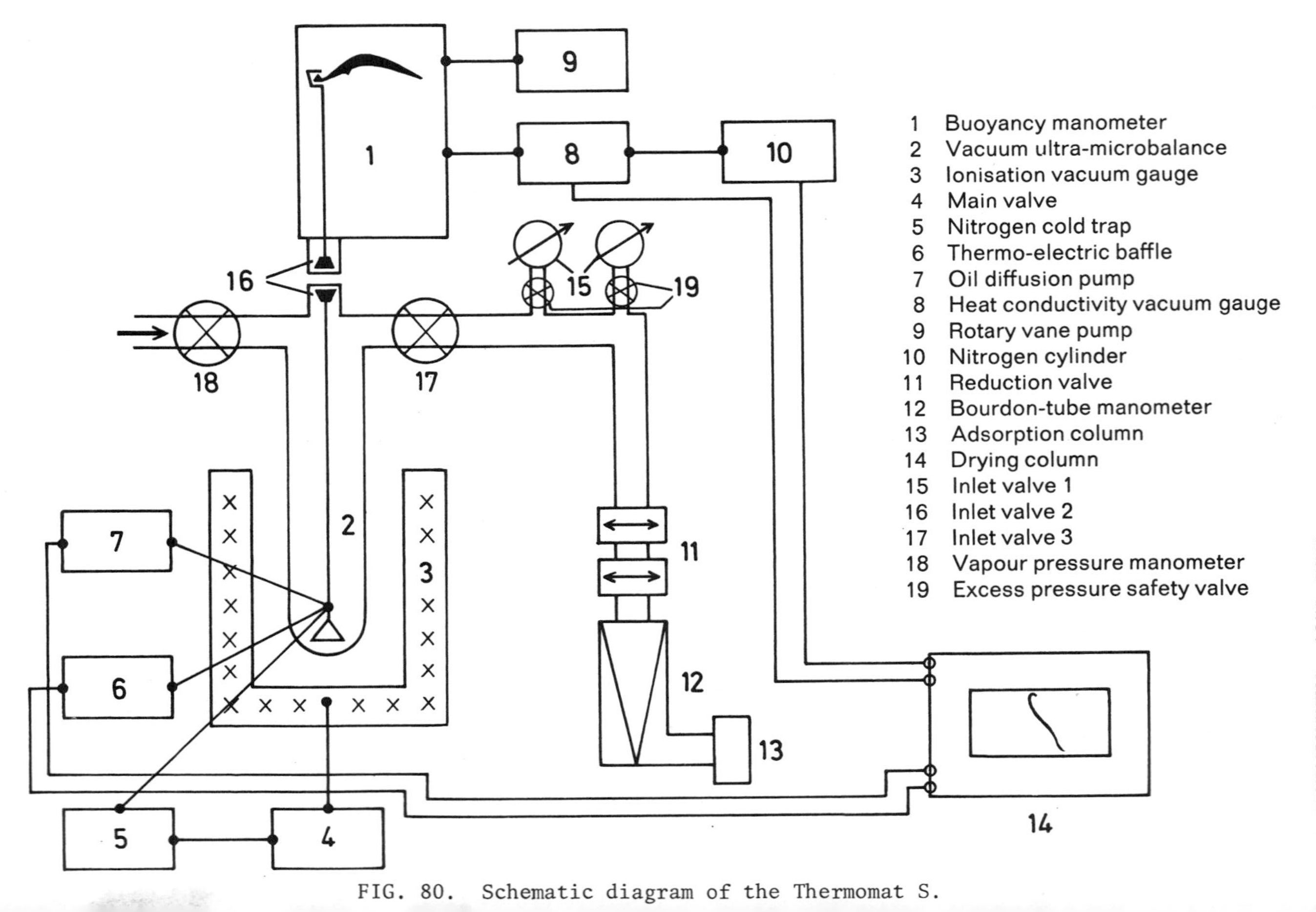
1 Buoyancy manometer
2 Vacuum ultra-microbalance
3 Ionisation vacuum gauge
4 Main valve
5 Nitrogen cold trap
6 Thermo-electric baffle
7 Oil diffusion pump
8 Heat conductivity vacuum gauge
9 Rotary vane pump
10 Nitrogen cylinder
11 Reduction valve
12 Bourdon-tube manometer
13 Adsorption column
14 Drying column
15 Inlet valve 1
16 Inlet valve 2
17 Inlet valve 3
18 Vapour pressure manometer
19 Excess pressure safety valve

FIG. 80. Schematic diagram of the Thermomat S.

Measuring Instruments

Penning vacuum gauge.

Measuring range: 10^{-2} to 10^{-6} Torr.

Quartz spring bourdon manometer.

Measuring range: 0 to 800 Torr.

4 channel recorder for weight and temperature recording.

Paper advance speed: 20-60-120-240-100-300-600-1200 mm/h.

Chart width: 250 mm.

Input voltage: 10 mV.

Accuracy of weight recording dependent on selected measuring range: ±0.01 mg to ±20 g.

Main voltage: 380 V, 50 Hz, 35 A.

Dimensions.

Width x height x depth: 155 x 188 x 65 cm.

Net weight: approx. 750 kg.

SETARAM

Perhaps the most elaborate TG systems available commercially are those manufactured by SETARAM (Societe d'Etude d'Automatisation de Regulation et d'Appareils de Measures). The philosophy of this complete line of thermal analysis equipment is to obtain the maximum amount of information from the sample with a single experiment. With this goal in mind, five different thermal analysis measurements are simultaneously carried out on a single sample; they are (a) TG, (b) DTG, (c) DTA, (d) analysis of gas emitted, and (e) temperature. Numerous accessories are available which permit studies to be made over a wide temperature range, in various gas atmospheres at different pressures, and in the presence of magnetic fields. Four types of furnace are employed, with maximum temperature ranges from 1000 to 2400°C; both atmospheric and vacuum versions of all four furnaces are available. As is the case with many other manufacturers, modular construction is used throughout the systems. Beginning with the base unit (represented by the B.70 balance and the measuring and control unit), the modular DTA unit, the derivative computer, and the gas analyzer can be connected easily. The cabinets are prewired and are of the plug-in type.

One of the basic balances used in the thermobalance system is the Ugine-Eyraud B.70 balance, illustrated in Fig. 81. The balance is of the beam type, equipped with three agate knife edges. The sample holder is suspended from the beam at one of the end knife edges. At the other end of the beam is the rebalancing suspension consisting of an optical shutter and a

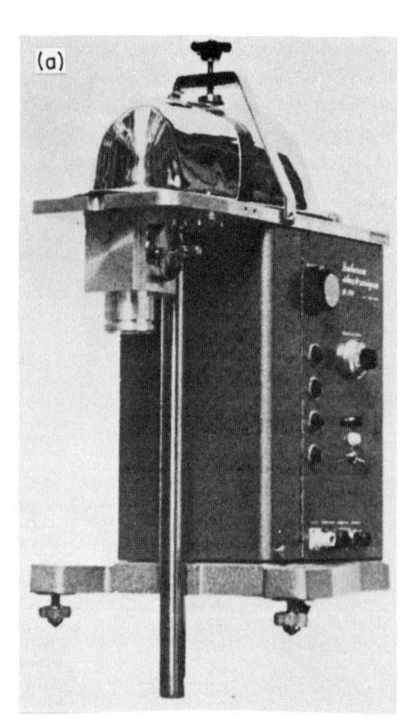

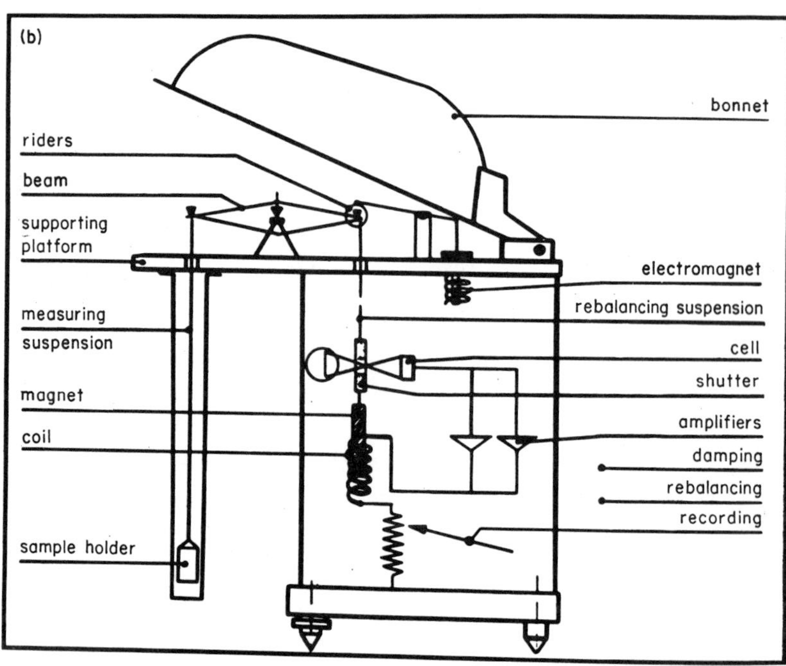

FIG. 81. SETARAM Ugine-Eyraud B.70 balance: (a) photograph, (b) schematic diagram.

permanent magnet. Suspension of the magnet in a solenoid coil enables the balance to be brought to an equilibrium position with respect to the sample mass charge. To supplement the range of automatic ranging, calibrated weights may be mechanically adjusted in steps of 10 to 150 mg and 100 to 1500 mg.

SPECIFICATIONS

Nominal load: 100 g.

Sensitivity: Without load - 1×10^{-5} g; with load - 1×10^{-5} g + 1×10^{-6} g per g of load.

Precision stability: 5×10^{-5} g.

Time constant of damping: 1 sec.

Measuring accuracy: 1×10^{-6}.

Mass ranges: Ten from 0 to 3.2 g: 75-150-300-600-1200 mg and 200-400-800-1600-3200 mg.

Circuits: Solid state.

Atmosphere: Low pressure to 5×10^{-6} Torr; pressure to 0.2 bar, relative.

The other balance that is used in the microthermoanalyzer series is the MTB 10-8 microbalance. This balance has the shape of a cylinder and is enclosed in a chamber 200 mm in diameter by 220 mm in height. The sealed enclosure contains the "weighing module" which consists of a symmetrical beam connected to a torsion band, a detector, and auxiliary units. A small diameter opening allows the introduction of a gas for atmosphere control. A "pumping block" permits the connection of the balance to pumping and vacuum measuring devices.

SPECIFICATIONS

Useful maximum load: 10 g (50 g optional).

Absolute sensitivity: 9.4 μg.

Absolute sensitivity: 4×10^{-8}.

Mass ranges: 0.05 to 20 mV/mg in 9 steps.

Measuring accuracy: 10^{-7}.

Types of operation:

controlled atmosphere,gas flow system, vacuum to 10^{-6} Torr.

There are four basic furnaces which are used in the SETARAM thermobalances with operating ranges from 25 to 2400°C. Specifications of these furnaces are given in Table 10.

TABLE 10

Specifications of SETARAM Furnaces

Parameter	1000° furnace	1600° furnace	2400° furnace	Micro-furnace
Maximum temperature, °C	1000	1600	2400 (in vac.)	1000
Heater element	Ni-Cr	W	C	Pt-Rh
Voltage, V	220	6		
Maximum power, kW	1.7	1.8	5	1.3 (low voltage)
Thermal inertia	1000° in 30 min.	1600° in 3 min.		
Cooling time, min	40			
Hot zone, mm	100	20	300	
Remarks		Operation under vac. (10^{-2} Torr) or gas flow. Vacuum version using C element also available.		Used mainly in magnetochemical studies.

FIG. 82. SETARAM MG 10S microthermoanalyzer.

Although it is not possible to discuss all of the thermobalances from this manufacturer, the Model MG 10S, which employs the MTB 10-8 balance, will be described briefly. This thermobalance is illustrated in Fig. 82.

The Model MG 10S system, which is the first of a series of six different thermobalances, includes the MTG 10-8 microbalance, a quartz hangdown tube, a 1000°C maximum temperature furnace, derivative computer for DTG plotting, a furnace programmer, and a six-channel potentiometric recorder. The balance mechanism is located on a floor mounted stand, while the other components are located in an enclosed cabinet. This system is for use at atmospheric pressure only.

The other thermobalances in this series are:

(a) MGVS 10S - identical to the MG 10S except for a high vacuum system.

(b) MG 17S - contains a 1700°C furnace and a vacuum system for 10^{-3} Torr operation.

(c) MGVS 17S - identical to the MG 17S except for addition of a high vacuum system.

Two other models are available, the MMD 11HT and MMD 11BT, which are used in magnetochemistry investigations.

The other series of thermobalances, The B-70 series, uses the Ugine Eyraud balance. They permit the simultaneous recording of the TG, DTG, DTA, EGD, and T curves of a single sample. The maximum temperatures of operation are 1000, 1600, or 2400°C.

STANTON REDCROFT

Stanton Redcroft Massflow and Standard Thermobalances

There are 14 models of the Stanton thermobalance, each differing in maximum temperature, heating rate, recording weight change, and so on. Specifications of the individual models are given in Table 11.

The massflow and standard thermobalances are shown in Fig. 83. Each thermobalance contains a precision air-damped analytical balance of proven design and a bi-filar wound tube-type furnace. The standard furnace is Nichrome wire-wound and is suitable for use to 1000°C; its internal bore is 2 in. in diameter. Alternative models are available with Pt/Rh wire windings for use to 1350 or 1550°C. All platinum-wound furnaces have an internal bore of 1.5 in. in diameter. The furnace is counterbalanced with its own low-voltage transformer and may easily be raised or lowered.

The sample is placed on a silica or alumina platform located in the furnace. The platform is mounted on a rod of similar refractory material rising from the top of the rear balance suspension.

Furnace temperature and change of weight are followed simultaneously on a two-pen recorder fitted above the balance and in front of the furnace. The curves are thus seen at eye level and are both shown by continuous lines side by side. The two pens are power driven by servomotors and receive their signals from a Pt, Pt-Rh thermocouple and a capacity follower plate located over the balance beam. This plate follows the

TABLE 11

Specifications of Stanton Thermobalances

	TR-1	TR-02	TR-01	HT-M	HT-F	HT-D	HT-SM	HT-SF	MF-L1	MF-L5	MF-H1	MF-H5	MF-S1	MF-S5
Max. temp., °C	1000	1000	1000	1400	1400	1400	1550	1550	1000	1000	1350	1350	1500	1500
Sensitivity, mg/div.	1.0	0.2	0.1	1.0	0.2	0.1	1.0	0.2	1.0	0.2	1.0	0.2	1.0	0.2
Max. linear heating rate	8	8	8	6	6	6	6.5	6.5	5.5	5.5	5.5	5.5	6	6
Chart range, mg	100	20	10	100	20	10	100	20	100	20	100	20	100	20
Automatic loading range, mg	±1000	200	100	1000	200	100	1000	200	1000	200	1000	200	1000	200
Max. rate of weight change	400	80	40	400	80	40	400	80	400	80	400	80	400	80
Sample capacity	20	20	20	16	16	16	16	16	20	20	20	20	20	20

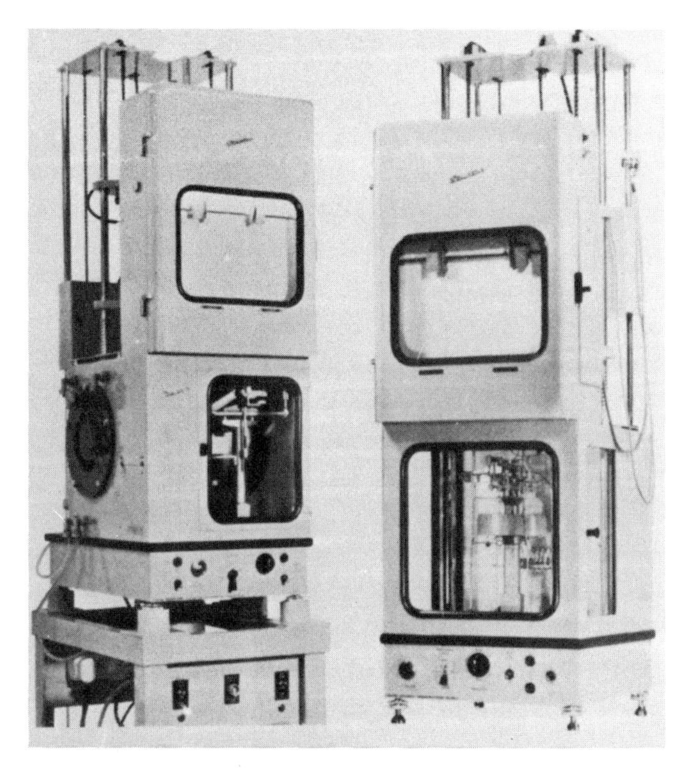

FIG. 83. Stanton thermobalances. Left: massflow model; Right: standard model.

movement of the beam yet has no direct or mechanical contact with it. Balance models have either 1 or 0.1 mg sensitivity. The servo-driven mechanism also operates an electric weight-loading device at the end of the beam, so that it is possible to follow weight changes up to ±1 g on the 1-mg sensitivity models or ±100 mg on the 0.1-mg models.

The balance is a constant load type; it is air-damped and equipped with synthetic sapphire knives and optically flat plates. The beam of the standard model is 12.5 cm in length, while the Massflow design has a 15-cm beam.

The recorder is a two-pen, 3-5 sec response time, using twin 12.5-cm chart paper scaled in mass and temperature. Chart speeds are 3, 6, and 12 in./hr.

The Standard Model reaction chamber can be operated only at atmospheric pressure. Greater variation in reaction chamber atmosphere can be made in the Massflow models: pressure may be varied from 10 psig to 10^{-4} Torr. Hydrogen can be used at high temperature only by special calibration.

Accessories include (a) alternative calibration device, (b) automatic cut-off switch, (c) thermostatically-controlled cold junction, (d) proportional temperature programmer, and (e) overhead pumping system for Massflow models.

DTA Attachment

Recently introduced is the DTA Attachment, which, when used in conjunction with the Standard Thermobalance, permits simultaneous DTA-TG measurements. It is possible to use the thermobalance alone or to use the DTA attachment independently. When used in the simultaneous mode, samples between 50 and 150 mg in weight are placed in the DTA crucibles. Thermogravimetry alone may be performed either in the DTA crucible or in a larger sample container mounted on the DTA block.

The DTA attachment is shown in Fig. 84. An alumina block, 20 mm in diameter and 13 mm in length, is used as the sample holder block. The block has two wells, each 6.5 x 10 mm, in which the Pt-Rh crucibles are placed.

A high-gain, low-noise, low-drift dc amplifier is used for the differential thermocouples. There is a four-step switch with ranges of 0.16 to 1.60°C/cm. Several optional recorders are available, such as the Leeds and Northrup Speedomax W single-pen (T and ΔT multiplexed on chart by a simple timer), a Leeds and Northrup X_1-X_2 recorder, and a modified Servoscribe RE520 X_1-X_2 recorder.

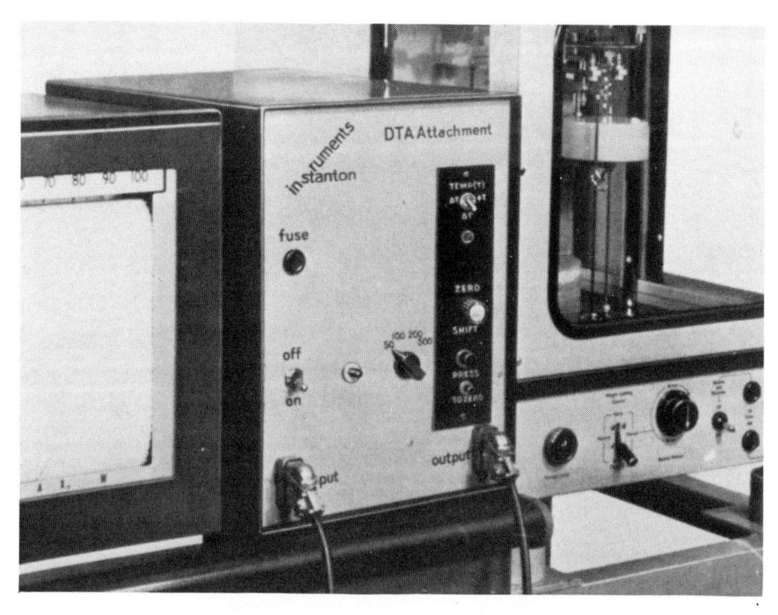

FIG. 84. Stanton DTA attachment for TG-DTA measurements.

Stanton Redcroft TG-750 Thermobalance

The TG-750 series thermobalance has been designed to give a direct plot of percentage mass change versus sample temperature, without the need for recalculating or replotting the results. It features a high-sensitivity electronic microbalance coupled to a water-cooled furnace which operates in the temperature range from 25 to 1000°C. The furnace is controlled by a modified Stanton Redcroft 681 Programmer using a platinum resistance thermometer as the sensor giving switch selected heating rates from 1 to 100°C/min. The design of the furnace permits rapid cooling (1000 to 100°C in about 4 min) and also enables selected isothermal temperatures to be established in a matter of seconds.

The electronic microbalance gives a range of switch selected sensitivities from 1 to 250 mg full-scale deflection. Because

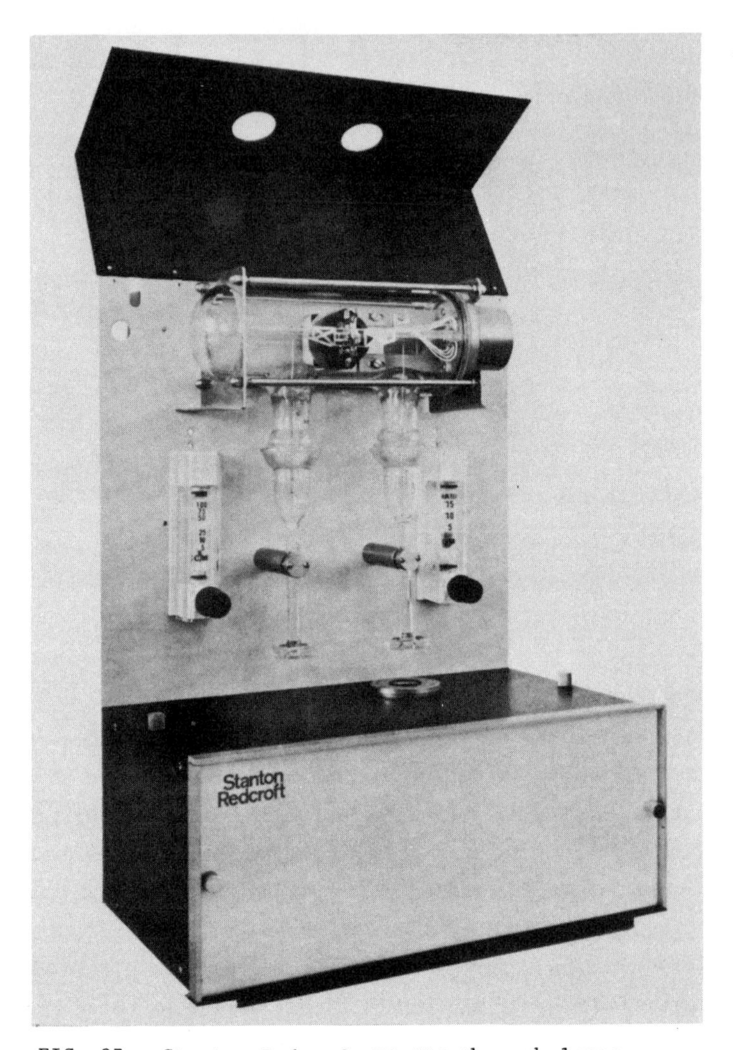

FIG. 85. Stanton Redcroft TG-750 thermobalance.

of the furnace design, there is said to be no noticeable buoyancy effect over the entire temperature range even at maximum sensitivity. It is, therefore, possible to record directly the percentage mass change of the sample as a function of the sample temperature. The unit can also plot the mass change derivative curve. Provision is made for atmosphere control and low pressure operation, and the small gas-swept volume of the furnace makes it suitable for connection to gas analysis systems including mass spectrometry.

The thermobalance is illustrated in Fig. 85.

THETA INDUSTRIES

Gravitronic TGA/DTA Apparatus

This apparatus is one part of the Theta thermal analysis system which includes DTA (previously discussed on p. 137), dilatometer, viscosimeter, and so on. Two models are available, differing in their balance capacity and maximum furnace temperature. Each model is capable of simultaneously recording both the TG and DTA curves of a sample. The model TGA 1a is illustrated in Fig. 86.

SPECIFICATIONS

	TGA 1a	TGA/DTA 111d
Temperature operating range	1000°C (-150° optional)	1750°C
Temperature programming	10 steps from 0.1 to 100°C/min	Same
Recorder	X-Y recorder, 22 x 28 cm	3 channel or multi-channel, 25 cm wide
Measuring system	Fused silica	Alumina
Capacity	2.15 g	100 g
Tare	2.50 g	100 g
Accuracy	$\pm 5 \times 10^{-4}$ of range	Same
Dimensions	65 x 150 cm (table mounted)	70 x 150 cm (floor mounted)
Power requirements	110 or 220 V ac; 1200 W	220 V ac; 3000 W

FIG. 86. Gravitronic TGA/DTA apparatus, Model TGA 1a.

VOLAND CORPORATION

$TRDA_3$-L and $TRDA_3$-H Derivative Thermobalances

There are two models of this thermobalance, as shown in Fig. 87, which simultaneously record the TG, DTG, and DTA curves of a sample.

The balance mechanism consists of a titanium beam with ruby knives and sapphire bearings. It is mounted in a stainless steel vacuum chamber with a single sealed control knob. The balance is arranged as a null-type, force-balance system, wherein the weight of the sample is detected and a restoring force in the form of an electric current applied to rebalance the beam. A portion of the balancing current is measured to indicate sample weight. All parts of the balance mechanism are fabricated from stainless steel, permitting its use with corrosive gases. The sample is suspended from a cantilevered support and is displaced horizontally from the balance, permitting the flow of gases over the sample without introducing vertical forces which would lead to incorrect mass indication.

The furnace of a horizontal tube type, enclosed in a vacuum envelope, which permits measurements to be made under vacuum or with flowing gases, is mounted on a slide to permit its removal from the sample holder. A preheater is enclosed within the furnace to permit gases to be brought to furnace temperature before exposure to the sample. Water cooling is also provided for rapid recycling. The standard furnace will maintain a temperature of 1000°C continously or 1200°C intermittently

FIG. 87. $TRDA_3$-L and $TRDA_3$-H derivative thermobalance.

($TRDA_3$-L); the high-temperature furnace will operate to a maximum temperature of 1500°C continuously or intermittently to 1600°C ($TRDA_3$-H).

An important feature of the sample holder is the method by which differential temperatures can be measured. The sample and reference materials are exposed to the same thermal conditions in that both are placed in shallow platinum cups which in turn are centered in platinum-rhodium thermal blocks mounted side by side on a ceramic tube. The sample holder is enclosed by a zero-gradient temperature zone within the furnace. Since the tube in horizontal, convection effects are minimized. Also, the shallow sample containers permit uniform temperatures through the sample material.

The temperature controller provides a trapezoidal cycle with a preselected heating rate. Temperature regulation is effected by a proportional, integral, and derivative (P.I.D.) control.

A solid-state potentiometric six point recorder is provided. Chart width is 180 mm, and printing time between points is 5 sec. Recording speeds from 50 to 600 mm/hr can be switch selected.

SPECIFICATIONS

<u>Balance</u>

Capacity: 1 g.

Sensitivity: 0.1 mg.

Material of beam: titanium.

Material of other parts: stainless steel.

Material of knife edges and bearings: ruby and sapphire.

<u>Furnaces</u>

$TRDA_3$-L: Iron-chromium-aluminum heater, noninductive winding.

Length: 360 mm.

Continuous temperature: Ambient - 1000°C.

Intermittent maximum temperature - 1200°C.

$TRDA_3$-H: Platinum-rhodium (10%) heater, noninductive winding.

Length: 360 mm.

Continuous temperature: Ambient - 1500°C.

Intermittent maximum temperature - 1600°C.

Amplifier/Ranges

Mass/Amplifier: 0-50 mg, 0-100 mg, 0-250 mg, 0-500 mg, 0-1000 mg (5 switched steps).

Differential/Thermal amplifier: 0-10 μV, 0-25 μV, 0-50 μV, 0-100 μV, 0-250 μV, 0-500 μV, 0-1000 μV (7 switched steps).

Mass differentiator: 1, 1/2.5, 1/5, 1/10 min^{-1} of mass range.

Temperature controller

Trapezoid P.I.D. 3 actions, S.C.R. continuous control.

Hold time: 0.5-24 hr continuously variable.

$TRDA_3$-L: 0.0-20°C/min (30 steps).
Maximum temperature: 1200°C.

$TRDA_3$-H: 0.14-26.7°C/min (30 steps).
Maximum temperature: 1600°C.

Recorder

Dot printing type Transistor type, recording potentiometer, 6 points.

Recording chart width: 180 mm.

Chart speed: 50, 100, 300, 600 mm/hr (4 switched steps).

$TRDA_3$-L: Recording range: 0-15 mV.

$TRDA_3$-H: Recording range: 0-20 mV.

TGA Module 1100-11

The TGA Module 1100-11, which is a plug-in module for the Voland Model 1100 thermalanalyzer system, utilizes the programmer, recorder, and furnace of the system. It consists of three parts: an electrical control console, a sample compartment and balance, and a stand. The module is connected to the Model 1100 unit by a cable.

There are three different models of balance available, each differing in their respective weighing ranges. One such balance and other parts of the module are illustrated in Fig. 88.

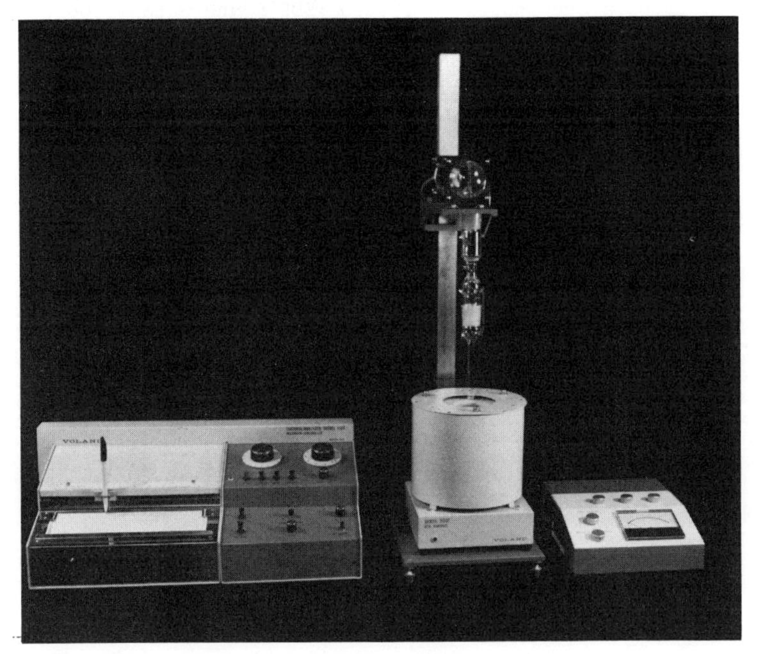

FIG. 88. Voland TGA Module 1100-11.

SPECIFICATIONS

Balance

Capacity: 1 g sample, 1 g counterweight

Models and weight ranges for the full 10-inch scale are as follows:

Range	Model 1100-11-A	Model 1100-11-B	Model 1100-11-C
1	0-25 μg	0-25 μg	0-1 mg
2	0-100 μg	0-250 μg	0-2.5 mg
3	0-250 μg	0-2.5 mg	0-10 mg
4	0-1 mg	0-10 mg	0-25 mg
5	0-2.5 mg	0-100 mg	0-100 mg

Temperature ranges: -100 to +500°C and 0 to +1000°C.

Electrical taring:

1. Coarse control - 9 steps of 1.1 mg
2. Medium control - 9 steps of 0.11 mg
3. Fine control - 1 mg total

Sensitivity: Approximately 4 x 10^{-3} full scale range (in milligrams).

Precision: ±5 x 10^{-7}

Pressure: atmospheric to 10^{-6} mm Hg.

Gas flow rate: 0 to 200 ml/min.

Sample pan size: 7 mm diameter.

Hangdown tubes: 13 mm outside diameter.

Heating rates: 0.1 to 12°C/min in 15 discrete steps.

WORDEN

Worden has developed a completely automatic microbalance which may be used for thermogravimetry. The Model 4302 is illustrated in Figs. 89(a) and 89(b).

The balance contains an all-fused quartz beam which is of the two-pan equal-arm design. Frictionless flexure hinges are used in place of knife edges or metallic taut-bands, which have friction or poor elastic properties. The readout is accomplished by using a pair of external torquing coils which produce a magnetic field of a permanent magnet mounted on the beam of the balance element. The position of the beam is monitored by a light source and photocell system, whose output is fed into a servocontroller. The servoloop is completed by connection across the torquing coils. Thus, when the photocells produce an error signal due to a mass charge, the servocontroller is activated and produces the required amount of current in the coils to bring the balance beam back to its null position. A precision microvolt meter completes the readout system by measuring the voltage drop across a temperature-controlled standard resistor. The readout is in digital form and reads directly in micrograms.

The balance is ideal for TG studies, because the balance chamber is designed for high-vacuum or controlled atmospheres. A special sample tube is available which allows the sample to be positioned in a furnace beneath the balance. The balance elements, being all fused quartz, are inert to most corrosive atmospheres. Available as an accessory is a 350-W, 110-V sample heater with a temperature controller. A complete TG console is available.

SPECIFICATIONS

Maximum sample weight: 10 g.
Minimum sample weight: 0.1 g.

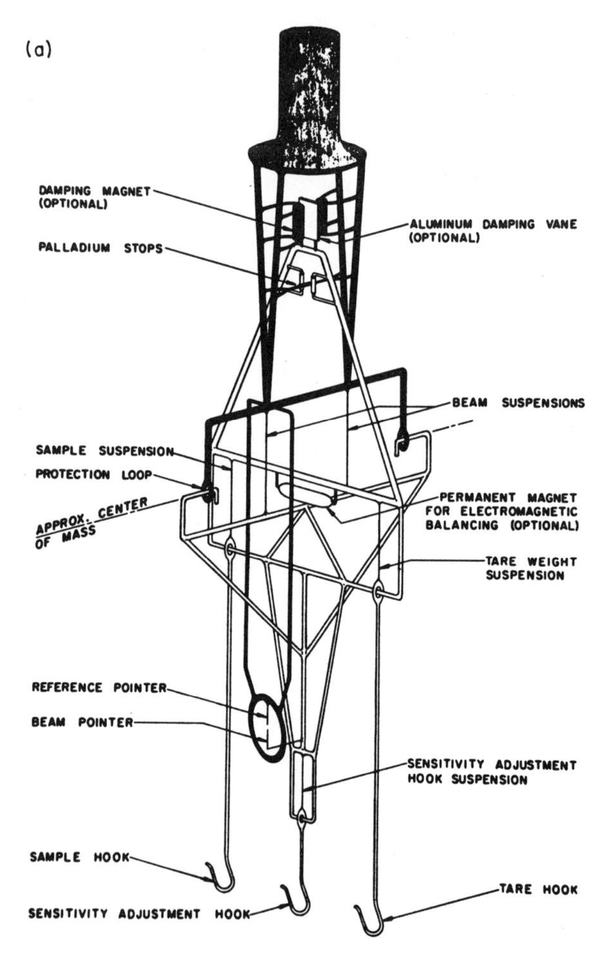

FIG. 89(a). Worden Model 4302 quartz microbalance: quartz beam.

FIG. 89(b). Worden Model 4302 quartz microbalance: photograph of balance.

Total delta weight range of servocontroller: 0-200 mg.

Resolution: 0.1 μg (1 part in 10^8 of maximum sample).

Precision (standard deviation): 0.7 μg.

Temperature coefficient: 1 μg/°C (under normal laboratory conditions).

Slew rate: 900 μg/sec maximum.

Calibration: Microvolt readout instrument may be calibrated by user to read directly in micrograms.

Mechanical sensitivity of balance element: Adjustable to infinity.

Readout: Manual null of microvolt instrument with digital display. Potentiometric recorder (option).

Recorder output: 0-0.25 V (adjustable).

Materials exposed in chamber: Pyrex, quartz, Viton, gold, and paladium.

Power requirements: 115 V 60 Hz (allowable fluctuations of ±10%).

APPENDIX: List of Manufacturers

A.D.A.M.E.L.
4-6 Passage Louis Phillipe
Paris (XIe)
FRANCE

Wm. Ainsworth and Sons, Inc.
2151 Lawrence Street
Denver, Colorado 80205

American Instrument Company
8030 Georgia Avenue
Silver Spring, Maryland 20910

Burrell Corporation
2223 Fifth Avenue
Pittsburgh, Pennsylvania 15219

Cahn Division
Ventron Instrument Corporation
7500 Jefferson Street
Paramount, California 90723

Columbia Scientific Industries
Analytical and Industrial Division
3625 Bluestein Boulevard
P. O. Box 6190
Austin, Texas 78762

DeltaTherm
Technical Equipment Corporation
917 Acoma Street
Denver, Colorado 80204

E. I. du Pont de Nemours and Company, Inc.
Instrument Products Division
Route 896
Glasgow, Delaware 19711

Eberbach Corporation
P. O. Box 63
Ann Arbor, Michigan 07481

Fisher Scientific
Instrument Division
711 Forbes Avenue
Pittsburgh, Pennsylvania 15219

Harrop Laboratories
3470 East Fifth Avenue
Columbus, Ohio 43210

Hungarian Optical Works
P. O. Box 52
Budapest 114
HUNGARY

Linseis
Vielitzer Strasse 43
8672 Selb
WEST GERMANY

Mettler Instrument Corporation
20 Nassau Street
Princeton, New Jersey 08540

Netzsch Geratebau GMBH
Werkstrasse 19
Selb/Bayern
WEST GERMANY

Edward Orton, Jr., Ceramic Foundation
1445 Summit Street
Columbus, Ohio 43210

Perkin-Elmer Corporation
Main Avenue
Norwalk, Connecticut 06852

Rigaku Denki Company, Ltd.
2-9-8 Sotokanda Chiyoda-ku
Tokyo
JAPAN

SETARAM
101-103 rue de Seze, 69
Lyon 6e
FRANCE

Shimadzu Seisakusho, Ltd.
14-5, Uchikanda 1-Chrome
Chiyoda-ku
Tokyo 101
JAPAN

Stanton Redcroft, Division of L. Oertling, Ltd.
Copper Mill Lane
London SW17 OBN
ENGLAND

Tem-Press Research
The Carborundum Company
1401 South Atherton Street
State College, Pennsylvania 16801

Tetrahedron Associates, Inc.
7605 Convoy Court
San Diego, California 92111

Theta Industries, Inc.
26 Valley Road
Port Washington, New York 11050

Voland Corporation
27 Centre Avenue
New Rochelle, New York 10801

Worden Quartz Products, Inc.
6121 Hillcroft Avenue
Houston, Texas 77036